VOLUME TWO HUNDRED AND TEN

METHODS IN CELL BIOLOGY

Cell Death – Part E

SERIES EDITOR

Lorenzo Galluzzi
Fox Chase Cancer Center,
Philadelphia, PA, United States

VOLUME TWO HUNDRED AND TEN

Methods in Cell BIOLOGY

Cell Death – Part E

Edited by

EMMA GUILBAUD
Fox Chase Cancer Center,
Philadelphia, PA, United States

RUTH SOLER-AGESTA
Weill Cornell Medical College,
New York, NY, United States

MANUEL BELTRÁN-VISIEDO
Fox Chase Cancer Center,
Philadelphia, PA, United States

LORENZO GALLUZZI
Fox Chase Cancer Center,
Philadelphia, PA, United States

ELSEVIER

ACADEMIC PRESS
An imprint of Elsevier

Academic Press is an imprint of Elsevier
125 London Wall, London, EC2Y 5AS, United Kingdom
50 Hampshire Street, 5th Floor, Cambridge, MA 02139, United States

First edition 2026

ISBN: 978-0-443-42955-2
ISSN: 0091-679X

For information on all Academic Press publications
visit our website at https://www.elsevier.com/books-and-journals

Publisher: Zoe Kruze
Editorial Project Manager: Devwart Chauhan
Production Project Manager: James Selvam
Cover Designer: Muthukumaran Thangaraj

Typeset by MPS Limited, India

Contents

Contributors

Clara Alcon
Department of Biomedical Sciences, Faculty of Medicine and Health Sciences, University of Barcelona; Cell Biology, Intracellular Compartments and Cancer, August Pi i Sunyer Biomedical Research Institute (IDIBAPS), Barcelona, Spain

Maria A. Bauer
Institute of Molecular Biosciences, NAWI Graz, University of Graz, Graz, Austria

Ingo Bechmann
Institute of Anatomy, University of Leipzig, Leipzig, Germany

Aziza Boyny
Chair of Microbiology, University of Würzburg, Würzburg, Bavaria, Germany

Daniel Butcher
Faculty of Health, Frazer Institute, Medicine and Behavioural Sciences, University of Queensland, Brisbane, QLD, Australia

Didac Carmona-Gutierrez
Institute of Molecular Biosciences, NAWI Graz; Field of Excellence BioHealth, University of Graz, Graz, Austria

Hojin Chang
Department of Biochemistry & Pharmacology & Biological Optical Microscopy Platform, The University of Melbourne, Parkville, VIC, Australia

Mathew J. Choaji
School of Biochemistry and Immunology, Trinity Biomedical Sciences Institute, Trinity College Dublin, Dublin, Ireland

Emma M. Creagh
School of Biochemistry and Immunology, Trinity Biomedical Sciences Institute, Trinity College Dublin, Dublin, Ireland

Faramarz Dehghani
Institute of Anatomy, University of Halle, Halle, Germany

Bahar Djouiai
Institute of Anatomy, University of Leipzig, Leipzig; GSI Helmholtzzentrum für Schwerionenforschung, Biophysics Department, Darmstadt, Germany

Felix Eichin
Institute for Developmental Immunology, Biocenter, Medical University of Innsbruck, Innsbruck, Austria

Tobias Eisenberg
Institute of Molecular Biosciences, NAWI Graz, University of Graz; BioTechMed-Graz, Graz, Austria; Field of Excellence BioHealth, University of Graz, Graz, Austria

Armin Frille
Division of Respiratory Medicine, Department of Medicine II, University of Leipzig Medical Center, Leipzig, Germany

Alejandra García-Carrasco
John B. Little Center for Radiation Sciences, Harvard T.H. Chan School of Public Health, Boston, MA, United States; Molecular and Integrative Physiological Sciences Program, Harvard T.H. Chan School of Public Health, Boston, MA, United States

Frank Gaunitz
Department of Neurosurgery, University Hospital Leipzig, Leipzig, Germany

Rüdiger Gerlach
Department of Neurosurgery, HELIOS Hospital Erfurt, Erfurt, Germany

Laura Hennig
Institute of Anatomy, University of Leipzig, Leipzig, Germany

Katharina Heß
Institute of Clinical Research and Systems Medicine, HMU Health and Medical University Potsdam, Potsdam, Germany

Rotraud Hirschberger
Institute for Developmental Immunology, Biocenter, Medical University of Innsbruck, Innsbruck, Austria

Ngoc Anh Hoang
Institute of Clinical Research and Systems Medicine, HMU Health and Medical University Potsdam, Potsdam; Division of Oncology, Department of Medicine II, University of Leipzig Medical Center, University Cancer Center Leipzig (UCCL), Leipzig, Germany

Albrecht Hoffmeister
Division of Gastroenterology, Department of Medicine II, University of Leipzig Medical Center, Leipzig, Germany

Katharina Kainz
Institute of Molecular Biosciences, NAWI Graz, University of Graz, Graz, Austria

Sonja Kallendrusch
Institute of Clinical Research and Systems Medicine, HMU Health and Medical University Potsdam, Potsdam; Institute of Anatomy, University of Leipzig, Leipzig, Germany

Rebecca Koch
Chair of Microbiology, University of Würzburg, Würzburg, Bavaria, Germany

Franziska Köhler
Institute of Anatomy, University of Leipzig, Leipzig; Department of Hematology and Oncology, Medical Faculty Mannheim, Heidelberg University, Mannheim, Germany

Florian Lordick
Division of Oncology, Department of Medicine II, University of Leipzig Medical Center, University Cancer Center Leipzig (UCCL), Leipzig, Germany

Frank Madeo
Institute of Molecular Biosciences, NAWI Graz, University of Graz; BioTechMed-Graz, Graz, Austria; Field of Excellence BioHealth, University of Graz, Graz, Austria

Joan Montero
Department of Biomedical Sciences, Faculty of Medicine and Health Sciences, University of Barcelona; Cell Biology, Intracellular Compartments and Cancer, August Pi i Sunyer Biomedical Research Institute (IDIBAPS), Barcelona; Networking Biomedical Research Center in Bioengineering, Biomaterials and Nanomedicine (CIBER-BBN), Madrid, Spain

Sam Nettelfield
Faculty of Health, Frazer Institute, Medicine and Behavioural Sciences, University of Queensland, Brisbane, QLD, Australia

Martin N. Odabas
Institute of Molecular Biosciences, NAWI Graz, University of Graz, Graz, Austria

Jiyi Pang
The Walter and Eliza Hall Institute of Medical Research; Department of Medical Biology, University of Melbourne, Parkville, Australia

Sebastian Prill
Division of Gastroenterology, Department of Medicine II, University of Leipzig Medical Center, Leipzig, Germany

Xingping Qin
John B. Little Center for Radiation Sciences, Harvard T.H. Chan School of Public Health; Molecular and Integrative Physiological Sciences Program, Harvard T.H. Chan School of Public Health; Laboratory of Systems Pharmacology, Harvard Program in Therapeutic Science, Department of Systems Biology, Harvard Medical School, Boston, MA, United States

Felicitas Rapp
Department of Hematology and Oncology, Medical Faculty Mannheim, Heidelberg University, Mannheim; GSI Helmholtzzentrum für Schwerionenforschung, Biophysics Department, Darmstadt, Germany

Steffen Rosahl
Department of Neurosurgery, HELIOS Hospital Erfurt, Erfurt, Germany

Thomas Rudel
Chair of Microbiology, University of Würzburg, Würzburg, Bavaria, Germany

Kristopher Sarosiek
John B. Little Center for Radiation Sciences, Harvard T.H. Chan School of Public Health; Molecular and Integrative Physiological Sciences Program, Harvard T.H. Chan School of Public Health; Laboratory of Systems Pharmacology, Harvard Program in Therapeutic Science, Department of Systems Biology, Harvard Medical School, Boston, MA, United States

Laura Schiller
Medical Faculty, Institute of Medical Physics and Biophysics, Leipzig University, Leipzig; Repha GmbH Biologische Arzneimittel, Langenhagen, Germany

Valentina C. Sladky
Institute for Developmental Immunology, Biocenter, Medical University of Innsbruck, Innsbruck, Austria

Aoife P. Smith
School of Biochemistry and Immunology, Trinity Biomedical Sciences Institute, Trinity College Dublin, Dublin, Ireland

Kathrin Stelzner
Chair of Microbiology, University of Würzburg, Würzburg, Bavaria, Germany

Andreas Villunger
Institute for Developmental Immunology, Biocenter, Medical University of Innsbruck, Innsbruck; The Research Center for Molecular Medicine (CeMM) of the Austrian Academy of Sciences, Vienna, Austria

James E. Vince
The Walter and Eliza Hall Institute of Medical Research; Department of Medical Biology, University of Melbourne, Parkville, Australia

Cica Vissiennon
Medical Faculty, Institute of Medical Physics and Biophysics, Leipzig University, Leipzig; Repha GmbH Biologische Arzneimittel, Langenhagen, Germany

Fabienne Wagner
Chair of Microbiology, University of Würzburg, Würzburg, Bavaria, Germany

Naiqi Wang
Faculty of Health, Frazer Institute, Medicine and Behavioural Sciences, University of Queensland, Brisbane, QLD, Australia

Julia Werner
Medical Faculty, Institute of Medical Physics and Biophysics, Leipzig University, Leipzig; Repha GmbH Biologische Arzneimittel, Langenhagen, Germany

Lachlan Whitehead
The Walter and Eliza Hall Institute of Medical Research; Department of Medical Biology, University of Melbourne, Parkville, Australia

Karsten Winter
Institute of Anatomy, University of Leipzig, Leipzig, Germany

Hubert Wirtz
Division of Respiratory Medicine, Department of Medicine II, University of Leipzig Medical Center, Leipzig, Germany

Daichao Xu
Interdisciplinary Research Center on Biology and Chemistry, Shanghai Institute of Organic Chemistry, Chinese Academy of Sciences; State Key Laboratory of Chemical Biology, Shanghai Institute of Organic Chemistry, Chinese Academy of Sciences; Shanghai Key Laboratory of Aging Studies; Shanghai Academy of Natural Sciences (SANS), Fudan University, Shanghai, P.R. China

Atomu Yamaguchi
John B. Little Center for Radiation Sciences, Harvard T.H. Chan School of Public Health, Boston, MA, United States; Department of Rehabilitation Science, Kobe University Graduate School of Health Sciences, Kobe, Hyogo, Japan; Molecular and Integrative Physiological Sciences Program, Harvard T.H. Chan School of Public Health; Laboratory of Systems Pharmacology, Harvard Program in Therapeutic Science, Department of Systems Biology, Harvard Medical School, Boston, MA, United States

Di Yu
Faculty of Health, Frazer Institute; Faculty of Health, Ian Frazer Centre for Children's Immunotherapy Research, Children's Health Research Centre, Medicine and Behavioural Sciences, University of Queensland, Brisbane, QLD, Australia

Zilong Zheng
Interdisciplinary Research Center on Biology and Chemistry, Shanghai Institute of Organic Chemistry, Chinese Academy of Sciences, Shanghai, P.R. China

Andreas Zimmermann
Institute of Molecular Biosciences, NAWI Graz, University of Graz; BioTechMed-Graz, Graz, Austria; Field of Excellence BioHealth, University of Graz, Graz, Austria

CHAPTER ONE

Flow cytometry analysis of mitochondrial apoptotic priming with BH3 profiling on senescent cells

Clara Alcon[a,b,*] **and Joan Montero**[a,b,c,*]

[a]Department of Biomedical Sciences, Faculty of Medicine and Health Sciences, University of Barcelona, Barcelona, Spain
[b]Cell Biology, Intracellular Compartments and Cancer, August Pi i Sunyer Biomedical Research Institute (IDIBAPS), Barcelona, Spain
[c]Networking Biomedical Research Center in Bioengineering, Biomaterials and Nanomedicine (CIBER-BBN), Madrid, Spain
*Corresponding authors. e-mail address: clara.alcon@ub.edu; jmontero@ub.edu

Contents

Abstract

Cellular senescence is a stable form of cell cycle arrest triggered by diverse stressors such as telomere shortening, oncogene activation, DNA damage, and chemotherapy. While senescence acts as a tumor-suppressive mechanism, the long-term accumulation of senescent cells contributes to chronic inflammation, tissue dysfunction, and age-related diseases, largely through the senescence-associated secretory phenotype

Methods in Cell Biology, Volume 210
ISSN 0091-679X, https://doi.org/10.1016/bs.mcb.2026.05.011

(SASP). Senescent cells are characterized by increased expression of $p16^{INK4a}$ and $p21^{CIP1}$, enlarged morphology, and resistance to apoptosis. This apoptotic resistance is mainly driven by the upregulation of anti-apoptotic BCL-2 family of proteins, including BCL-2, BCL-xL, and MCL-1. Senolytics – drugs that selectively induce apoptosis in senescent cells – target these survival pathways to promote senescent cell clearance. BH3 profiling is a functional assay that assesses mitochondrial apoptotic priming by exposing permeabilized cells to BH3-domain peptides and measuring cytochrome c release. This technique allows the identification of anti-apoptotic dependencies in senescent cells and can be used in combination with senescence markers (p16 and p21) via flow cytometry to pinpoint populations with enhanced apoptotic resistance. Understanding the apoptotic landscape of senescent cells is critical for optimizing senolytic strategies and improving therapeutic outcomes, particularly in contexts such as cancer and therapy-induced senescence.

1. Introduction

Senescence is a state of stable cell cycle arrest triggered by various stressors, including telomere shortening, oncogene activation, DNA damage, oxidative stress, and chemotherapeutic agents (Di Micco et al., 2021; Yamauchi & Takahashi, 2025). Depending on the stressor, cellular senescence can be divided into oncogene-induced senescence (OIS), replicative senescence and therapy-induced senescence (TIS). Senescence serves as a potent tumor-suppressive mechanism by preventing the propagation of damaged cells, but the long-term accumulation of senescent cells can contribute to tissue dysfunction, inflammation, and age-related diseases (Di Micco et al., 2021). While senescent cells no longer proliferate, they remain metabolically active and undergo profound phenotypic changes. These include increased cell size, chromatin remodeling, elevated lysosomal activity, and the secretion of a diverse set of bioactive molecules collectively known as the senescence-associated secretory phenotype (SASP) (Kirichenko et al., 2025; Wang et al., 2024). Senescent cells' growth arrest is mediated through the activation of tumor suppressor pathways, primarily $p16^{INK4a}$/Rb and p53/p21. The overexpression of p16 and p21, together with the phenotypic changes and the senescence-associated β-galactosidase activity are typically used to identify senescent cells (González-Gualda et al., 2021).

Senescent cells are known to resist apoptosis despite being exposed to cellular stressors that would normally trigger programmed cell death. This resistance is largely due to the overexpression or overactivation of anti-apoptotic pathways, particularly involving members of the BCL-2 family of proteins such as BCL-2, BCL-xL and MCL-1. These proteins inhibit the activation of pro-apoptotic effectors like BAX and BAK, and prevent the activation of pro-apoptotic proteins such as BIM and BID, thereby preventing mitochondrial outer membrane permeabilization (MOMP) and the downstream cascade leading to cell death (Kale et al., 2018; Singh et al., 2019). The persistence of senescent cells in tissues contributes to chronic inflammation, tissue dysfunction, and age-related diseases, largely due to the SASP. To counteract these detrimental effects, a class of drugs called senolytics has been developed (Kirkland & Tchkonia, 2020). Senolytics are agents that selectively induce apoptosis in senescent cells by targeting their pro-survival mechanisms, particularly the BCL-2 family proteins or components of the PI3K/AKT and p53 pathways. Examples of senolytics include navitoclax (ABT-263), which inhibits BCL-2/BCL-xL, and combinations like dasatinib and quercetin, which target multiple survival pathways (Maurer et al., 2025; Zhu et al., 2016). By clearing senescent cells, senolytics hold therapeutic promise for improving health span, reducing inflammation, and enhancing the efficacy of cancer therapies (Galiana et al., 2020; Ogrodnik et al., 2021; Xu et al., 2018; Zhang et al., 2022).

Given that the proportion of cells entering senescence during treatment can vary depending on the cell line and the senescence-inducing agent, accurately identifying this population is essential for detecting apoptosis-related changes. To this end, evaluating the BCL-2 family of proteins specifically in senescent cells would be crucial to study the apoptotic status of these cells and identify possible anti-apoptotic vulnerabilities to target with BH3 mimetics. BH3 profiling is a functional assay that measures mitochondrial apoptotic priming by assessing the cell's susceptibility to undergo apoptosis via the intrinsic pathway (Certo et al., 2006; Ryan & Letai, 2013). This technique involves exposing permeabilized cells to synthetic ~ 20-mer BH3 peptides derived from BH3-only proteins, some of which antagonize anti-apoptotic BCL-2 family members and activate pro-apoptotic effectors like BAX and BAK. The resulting MOMP is evaluated through cytochrome c release typically measured by flow cytometry (Chonghaile et al., 2011; Ryan et al., 2016). By using the BIM peptide, BH3 profiling provides a quantitative readout of how close a cell is to the apoptotic threshold, known as 'overall priming', offering valuable insights into treatment resistance, particularly in contexts such as cancer and therapy-induced senescence. Furthermore, using sensitizer BH3 peptides which have specific affinity for anti-apoptotic proteins (BAD for BCL-2, BCL-xL and BCL-w; HRK for BCL-xL and MS1 for MCL-1) we can identify apoptotic resistance mechanisms driven by a specific anti-apoptotic protein (Montero & Letai, 2018). The identification of the anti-apoptotic vulnerabilities of a senescent population of cells is crucial to use BH3 mimetics as senolytics (Chang et al., 2016; Yang et al., 2025; Yosef et al., 2016; Zhu et al., 2015). Flow cytometric analysis of cytochrome c retention in mitochondria, combined with the use of p16 and p21 as senescence markers, enables the discrimination of cell populations that are more resistant to apoptosis.

2. Materials

2.1 Common disposables

- TPP® 15 mL centrifuge tubes (#Z707724, TPP)
- Tissue Culture Flasks 25–300 cm^2 (#90025, TPP)
- Eppendorf™ Safe-Lock Tubes 1.5 mL – Microtube (#15625367, Thermo Fisher)
- TPP® tissue culture plates (#92006, TPP)
- Corning® 96 Well Clear Polystyrene Microplate, round bottom (# CLS3795, Corning)

2.2 Reagents

- PE Waf1/Cip1/CDKN1A p21 antibody (F-5) (#sc-6246, Santa Cruz Biotechnology)
- Alexa Fluor® 488 CDKN2A/p16INK4a (# bs-4592R-A488, Bioss Antibodies)
- Alexa Fluor® 647 anti-Cytochrome c—6H2. B4 (#612310, BioLegend)
- Gibco™ RPMI 1640 (#31870, Thermo Fisher)
- Gibco™ Trypsin-EDTA (0.05 %), phenol red (#11580626, Thermo Fisher)
- Gibco™ Penicillin-Streptomycin (10,000 U/mL) (#11568876, Thermo Fisher)
- Gibco™ L-Glutamine (200 mM) (#11534546, Thermo Fisher)
- FBS Xtra, Fetal Bovine Serum (FBS), Collected in South America (# FBS-16A, Labclinics)
- Zombie Violet™ Fixable Viability Kit (#423113, BioLegend)
- Mannitol (#M4125-100G, Sigma-Aldrich)
- HEPES (#H4034, Sigma-Aldrich)
- EDTA (#E6758, Sigma-Aldrich)

- EGTA (#E3889, Sigma-Aldrich)
- Bovine serum albumin (#A3059, Sigma-Aldrich)
- Succinic Acid (#S3674, Sigma-Aldrich)
- Potassium Chloride (#P9541, Sigma-Aldrich)
- Digitonin (#D5628, Sigma-Aldrich)
- Alamethicin (#HY-N6708, MedChemExpress)
- Potassium Hydroxide (#P5958, Sigma-Aldrich)
- Tris base (#10376743, Thermo Fisher)
- Glycine (#5046, Sigma-Aldrich)
- Formaldehyde solution (#F1635, Thermo Fisher)
- BH3 peptides:

Peptide	Sequence
hBIM	Acetyl-MRPEIWIAQELRRIGDEFNA-Amide
Hrk-y	Acetyl -SSAAQLTAARLKALGDELHQY- Amide
MS1	Acetyl-RPEIWMTQGLRRLGDEINAYYAR-Amide
mBAD	Acetyl -LWAAQRYGRELRRMSDEFEGSFKGL- Amide

2.3 Equipment

- CytoFLEX flow cytometer (Beckman Coulter Life Sciences)
- Benchtop centrifuge Gyrozen (#1248R, Controltecnica)
- Benchtop microfuge Gyrozen (#1730R, Controltecnica)
- Humidified cell culture incubator (Heracell™ 150i CO2, Thermo Fisher)
- Laboratory biosafety cabinet (Class II, MARS XXX, Controltecnica)

2.4 Software

- FlowJo v.10.9.0 (FlowJo LLC)
- Excel (Microsoft)
- Prism v 10.0.1 (GraphPad)

3. Methods

3.1 Therapy-induced senescence

1. Tumoral cells are routinely expanded at 37 °C and 5 % CO_2 in Tissue Culture Flasks 25–300 cm^2 containing 10 mL RPMI1640 supplemented with 10 % heat-inactivated FBS, 1% L-glutamine and 1% penicillin and streptomycin.

2. When tumoral cells reach 80–90 % confluence, cells were counted and seeded in TPP® 6 well tissue culture plates (See Note 1).
3. Cells were treated with 1 μM palbociclib and after 3 days, cells were replenished with fresh media and 1 μM palbociclib.
4. After 7 days, cells are ready to perform BH3 profiling (Fig. 1).

3.2 BH3 profiling

1. We perform BH3 profiling as described by the Letai lab (Ryan et al., 2016). Prepare the different dilutions of the BIM peptide (0.01 μM, 0.03 μM, 0.1 μM, 0.3 μM, 1 μM, 3 μM and 10 μM), BAD (10 μM), HRK (100 μM), MS1 (10 μM) and alamethicin (50 μM) in MEB (150 mM mannitol, 10 mM HEPES-KOH pH 7.5, 150 mM KCL, 1 mM EGTA, 1 mM EDTA, 0.1 % BSA, 5 mM succinate) buffer and 0.001 % digitonin (See Note 2 and 3).
2. Place the different BIM concentrations, sensitizer peptides (BAD, HRK and MS1) alamethicin and a negative control (MEB buffer + digitonin) on a 96 well plate. We will use 25 μLof each peptide concentration per experimental condition (See Note 4).
3. Trypsinize cells and centrifuge them to remove the supernatant.
4. Stain cells with zombie violet (1:100 dilution in PBS) for 10 min at RT.
5. Centrifuge cells and wash the pellet with PBS.
6. Centrifuge cells and resuspend the pellet in MEB buffer (See Note 5).
7. Place 25 μL of cells in the 96 well plate containing the different peptides and incubate 1 h at RT. (See Note 6)
8. Fix cells by directly adding 25 μL of formaldehyde 8 % in each well that already contains 25 μl of peptide solution and 25 μL of cell suspension.

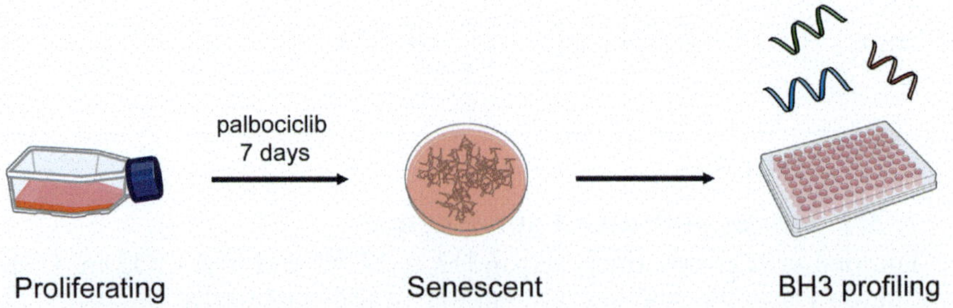

Fig. 1 Experimental design. Proliferating cells will be treated with palbociclib 1 μM for 7 days, replenishing the media at day 3 and adding 1 μM palbociclib. BH3 profiling will be performed at day 7 in proliferating and senescent cells.

9. After 15 min, neutralization of formaldehyde is performed by adding 50 μL of N2 buffer (1,7 M Tris base, 1.25M Glycine, pH 9.1) into each well.

3.3 Staining

1. Cells are stained with cytochrome c, p21-PE and p16-A488 in intracellular buffer (1 % Tween20, 5 % BSA in PBS) and plates are left overnight at 4 °C (See Note 7). It is important to have a negative control (without antibody for p16 and p21) to prove for correct staining (See Note 8 and 9).

3.4 Acquisition and gating

1. Laser gain and compensation are set on unstained cells as well as cells stained with each fluorochrome separately.
2. Acquire 3000 events per well (See Note 10).
3. Gating should be performed as followed (Fig. 2):
   a. Cells are identified using FSC-A and SSC-A dot plots. Events with low FSC or SSC are excluded as they are probably debris. Proliferating and senescent cell gates may be different as senescent cells become more complex and bigger in size (Fig. 2A) (See Note 11).
   b. Alive cells are identified using Zombie Violet™ (V450 laser). Unstained events are the alive population of cells (Fig. 2B).
   c. On alive cells, senescent cells are identified on FSC-A vs PE (B585 laser) or A488 (B525 laser) (See Note 12). Always use the negative control (without antibody) to define p21 and p16 positive staining (Explained in Note 8) (Fig. 3).

   Do a gating using proliferating and senescent cells as p21/p16 high (senescent) and p21/p16 low (proliferating) (Fig. 4) (See Note 9).

   d. Induction of apoptosis is then assessed in the proliferating (PE/A488 low) and in the senescent population (PE or A488 high) by using FSC-A vs A647 (R660 laser). DMSO and alamethicin are used to define the cytochrome c retained inside the mitochondria (Fig. 5) (See Note 13). Cytochrome c retention gate should be adjusted depending if cells are proliferating or senescent (See Note 14) (Fig. 6). Fig. 7

3.5 Analysis

1. With the % cytochrome c retained in proliferating or senescent alive cells acquired from the flow cytometry analysis, we obtain the dose response curves for the different concentrations of the BIM

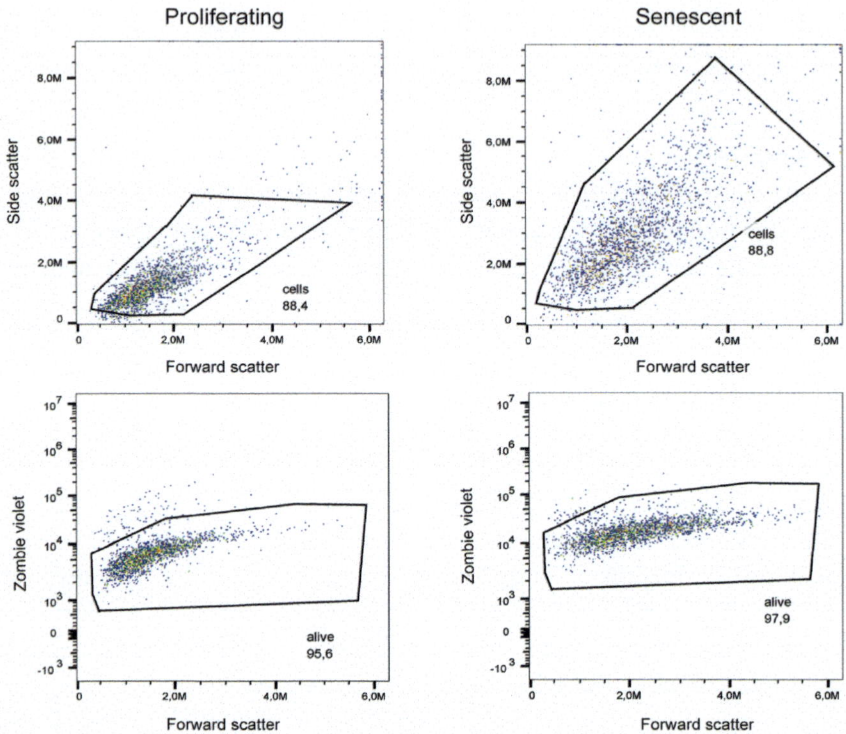

Fig. 2 Gating strategy. (A) Cell population gating for proliferating and senescent cells. Senescent cells exhibit increased size and complexity; therefore, the gating strategy must be adjusted in the FSC and SSC plots for each experimental condition. (B) Zombie Violet gating. Cells negative for Zombie Violet staining represent the viable cell population.

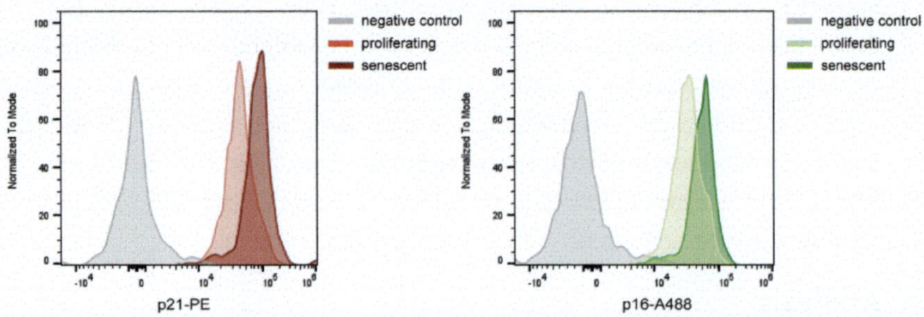

Fig. 3 p21 and p16 plots for negative control (cells without antibodies), proliferating cells and senescent cells. There is an increase in intensity in senescent cells compared to proliferating cells.

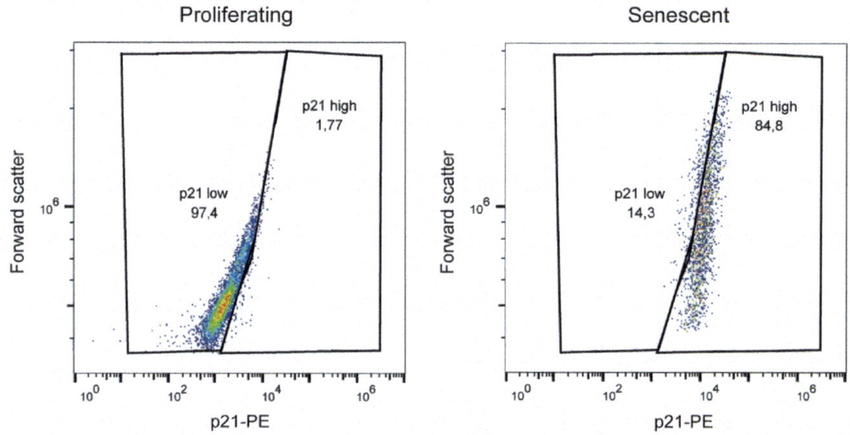

Fig. 4 p21/p16-high and low gating strategy. As proliferating cells also express p21 and p16, low and high gates are used to identify non-senescent and senescent cells.

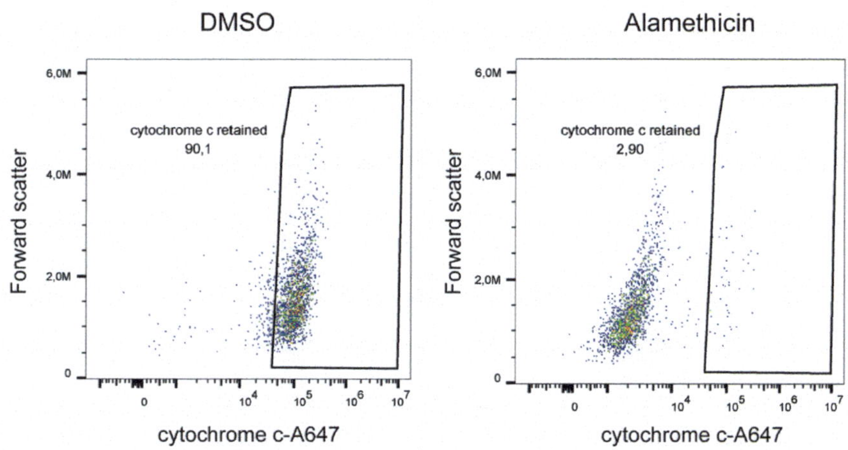

Fig. 5 Cytochrome c retention gating strategy. DMSO and alamethicin treated samples are used to define the gates for cytochrome c retention within mitochondria. Cells treated with alamethicin exhibited complete cytochrome c release, whereas DMSO-treated cells retained cytochrome c.

peptide (Fig. 8A) (See Note 15). For the sensitizer peptides we follow the same strategy only using one concentration of each peptide (Fig. 8C) (See Note 16)

2. We then calculate the cytochrome c released from inside the mitochondria (100 - cytochrome c retained).

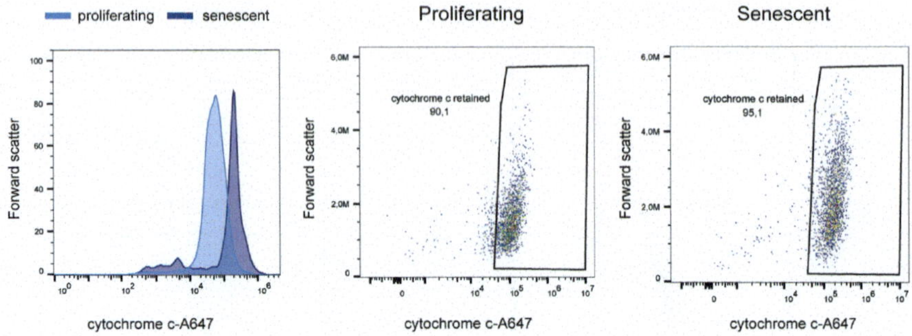

Fig. 6 Gating adjustment of cytochrome c in proliferating vs senescent cells. Senescent cells possess enlarged mitochondria and consequently higher cytochrome c content. Therefore, the cytochrome c gate must be adjusted in senescent cells to ensure accurate analysis of changes detected during BH3 profiling.

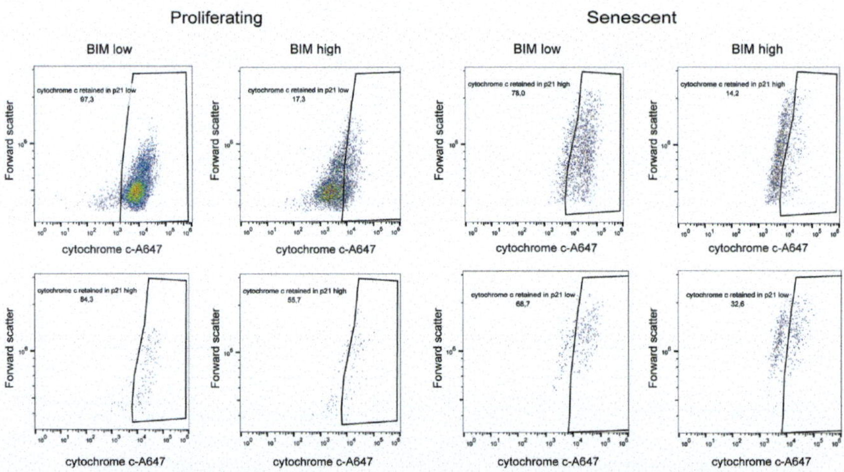

Fig. 7 Cytochrome c released from inside the mitochondria in p21 high or p21 low population of cells after exposing them to a low or high concentration of the BIM peptide.

3. Δ% priming is then calculated by subtracting the cytochrome c released in the experimental condition - the cytochrome c released in the control condition for every BIM concentration (Fig. 8B) or for the sensitizer peptides (Fig. 8D) (See Note 17).

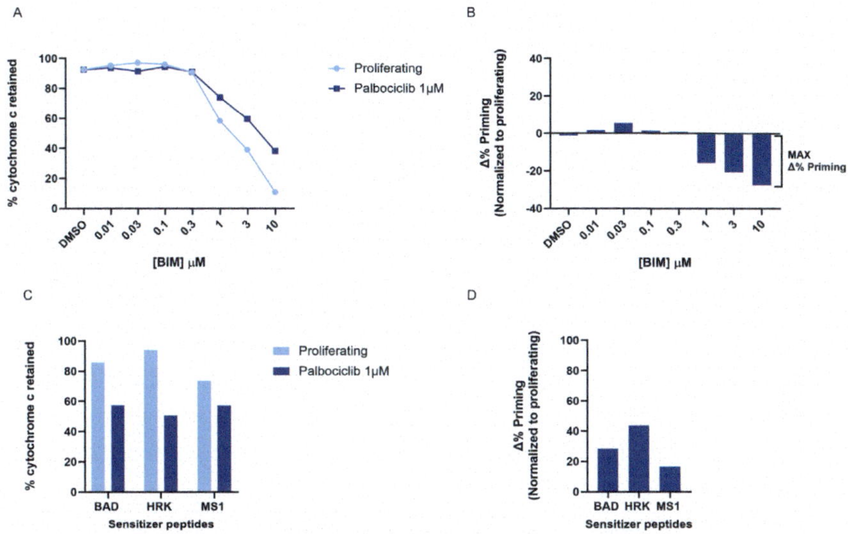

Fig. 8 Examples of obtained plots after BH3 profiling. (A) Cytochrome c retention plot with BIM peptide in proliferating and palbociclib – induced senescent cells. (B) Δ% Priming of palbociclib-treated cells compared to proliferating cells. Negative Δ% priming values at higher BIM concentrations reflect the resistance of palbociclib-induced senescent cells to apoptosis. (C) Cytochrome c retention plot with sensitizer peptides in proliferating and palbociclib – induced senescent cells. (D) Δ% Priming of palbociclib-treated cells compared to proliferating cells. Positive Δ% priming indicates that senescent cells have acquired dependence on the anti-apoptotic proteins targeted by the sensitizer peptides. In this example, the strongest dependency is observed with the HRK peptide, suggesting an adaptive reliance on BCL-xL.

4. Notes

1. On the day of seeding, it is important to consider that proliferating - but not senescent -cells will continue to divide during the following seven days. Therefore, proliferating and treated cells should be seeded at different densities. Pilot experiments are recommended to determine the appropriate number of cells to seed for each condition. When performing BH3 profiling (day 7), 300,000 cells per experimental condition are required.
2. The titration of the BIM peptide should be optimized for each cell type, typically ranging from 0.01 μM to 10 μM. However, for more resistant cell lines, concentrations up to 100 μM of BIM may be

necessary. When calculating the BIM peptide concentrations, we must consider that these solutions will be diluted; therefore, we need to prepare them at 2 × concentration.

3. Digitonin is a mild detergent composed of a steroid head and a polysaccharide tail that forms insoluble complexes with plasma membrane cholesterol. Its biological activity therefore depends on the cholesterol content of the cell membrane, and the optimal digitonin concentration may vary between cell types. Therefore, adjusting the concentration of digitonin could be necessary in some experiments.
4. The total volume of MEB buffer and digitonin needed for the preparation of the different BIM peptide concentrations should be calculated as 25 μL per well × number of experimental conditions × number of BH3 profiling conditions (BIM titration, sensitizer peptides, alamethicin and DMSO).
5. The MEB buffer used to resuspend cells should not contain digitonin, and the final resuspension volume will depend on the number of conditions tested. For example, if the following conditions are included: DMSO, alamethicin, BIM 0.01 μM, BIM 0.03 μM, BIM 0.1 μM, BIM 0.3 μM, BIM 1 μM, BIM 3 μM, and BIM 10 μM, the required total volume will be 25 μL × 9 = 225 μL.
6. Mitochondrial depolarization is highly influenced by temperature. The incubation of the cells with the peptides should be performed always at a constant temperature. If the temperature in the lab shifts between seasons it will be better to incubate the cells in an incubator with a controlled temperature.
7. It is not necessary to analyze the samples immediately after staining. The 96-well plates can be stored at 4 °C for several weeks prior to flow cytometry analysis.
8. Include one additional well for the DMSO condition and incubate these cells with intracellular buffer without p21/p16 antibodies to serve as a negative control for p21 and p16 staining verification.
9. As p16 and p21 are proteins expressed in both proliferating and senescent cells, high p21/p16 and low p21/p16 staining will be use to separate proliferating and senescent cells.
10. Acquiring a moderate number of events per sample is essential to ensure reliable gating and accurate data analysis. If there are 300.000 cells per experimental condition at the starting point, after splitting the samples into 12 wells, it should be around 25.000 cells/well.

11. Senescent cells are larger and more complex than proliferating cells. As FSC and SSC parameters may differ between conditions, the cell gate must be adjusted according to the experimental group (proliferating vs. senescent) (See Fig. 2).
12. It is important to stain cells for both p16 and p21, since not all senescent cells upregulate both markers. Some therapy-induced senescent cells may show increased expression of only one of them. Assessing both markers ensures accurate identification of the senescent population.
13. DMSO and alamethicin conditions are used to define the gating for cytochrome c retention inside the mitochondria. Cells treated with alamethicin should exhibit complete cytochrome c release, whereas DMSO treated cells should have it retained. These two conditions are therefore used to set the retention gate (See Fig. 5).
14. Senescent cells typically display enlarged mitochondria and higher cytochrome c content. Consequently, cytochrome c staining differs between proliferating and senescent cells, and the cytochrome c retention gate should be adjusted accordingly for each condition (See Fig. 6).
15. For proliferating cells, the cytochrome c retention analysis should be performed on the population of viable cells. In the case of senescent cells, cytochrome c retention should be analyzed in viable p21 high or p16 high cells.
16. Concentrations of the sensitizer peptides should be adjusted depending on the cell type. For example, cells from hematological tumors have high dependence on BCL-2, therefore, BAD and HRK peptide concentrations should be reduced. Therefore, pilot experiments should be performed to decide the optimal concentration of the sensitizer peptides. Sometimes, performing a titration similar to BIM peptide is needed.
17. Senescent cells are resistant to enter into apoptosis, therefore, the Δ% priming obtained in this population of cells could be negative.

5. Concluding remarks

The protocol described here outlines an approach to evaluate the apoptotic predisposition of therapy-induced senescent cells. BH3 profiling enables the assessment of apoptotic priming in a cell population by measuring mitochondrial outer membrane permeabilization (MOMP) (Certo et al., 2006; Ryan & Letai, 2013).

A defining feature of senescent cells is their resistance to apoptosis, often driven by the overexpression or overactivation of anti-apoptotic proteins (Barriuso et al., 2023; Jia et al., 2023; Yosef et al., 2016). Identifying these anti-apoptotic vulnerabilities is critical for eliminating senescent cells through BH3 mimetics or alternative senolytic strategies such as PROTACs (Selt et al., 2023; Shahbandi et al., 2020). By using specific peptides that target anti-apoptotic proteins, BH3 profiling enables the identification of these dependencies. This approach was recently applied to characterize anti-apoptotic vulnerabilities in therapy-induced senescent melanoma cells following treatment with palbociclib and irradiation, where inhibition of BCL-xL successfully triggered apoptosis (Alcon et al., 2025).

Importantly, BH3 profiling not only identifies apoptotic priming and resistance mechanisms but also helps define the senescent cell fraction within a treated population - which is often less than 100%. This allows for a more precise determination of anti-apoptotic dependencies that can subsequently be targeted with BH3 mimetics or PROTACs.

Another major advantage of BH3 profiling is its ability to generate predictive data within a short timeframe, which is particularly valuable when analyzing primary samples that rapidly lose viability ex vivo. Therefore, assessing apoptotic priming and anti-apoptotic dependencies in patient-derived samples can have an impact in detecting senescent cells and in designing targeted therapeutic strategies to eliminate them.

Acknowledgements

We would like to thank the Cytometry Facility, the Irradiation Service and the Electron Microscopy Unit from the Scientific and Technological Centers of the University of Barcelona (CCITUB) for their assistance with experiments. We would like to acknowledge Ministerio de Ciencia, Innovación y Universidades Grant RTI2018-094533-A-I00, PID2021-128644OB-I00 and PID2024-159425OB-I00 funded by MICIU/AEI/ 10.13039/501100011033 and by "ERDF A way of making Europe" grants; CELLEX foundation; José Carreras Leukämie-Stiftung grant DJCLS 19R/2022; and Agència de Gestió d'Ajuts Universitaris i de Recerca-Generalitat de Catalunya 2021-SGR-00284, FP00019758_SA001 (Prime agreement: 2R01CA215610-05, National Institutes of Health), V beca trienal Fundación Mari Paz Jiménez Casado and ICREA Acadèmia 2024.

Competing interests

JM is co-inventor of dynamic BH3 profiling (patented by Dana-Faber Cancer Institute) and has received royalties. JM is a paid consultant for PEP-THERAPY, was a paid consultant for Oncoheroes Biosciences and Vivid Biosciences, is an unpaid board member for The Society for Functional Precision Medicine, and he is currently collaborating with AstraZeneca.

References

Alcon, C., Kovatcheva, M., Morales-Sánchez, P., López-Polo, V., Torres, T., Puig, S., ... Montero, J. (2025). HRK downregulation and augmented BCL-xL binding to BAK confer apoptotic protection to therapy-induced senescent melanoma cells. *Cell Death & Differentiation, 32*(4), 646–656. https://doi.org/10.1038/s41418-024-01417-z.

Barriuso, D., Alvarez-Frutos, L., Gonzalez-Gutierrez, L., Motiño, O., Kroemer, G., Palacios-Ramirez, R., & Senovilla, L. (2023). Involvement of Bcl-2 family proteins in tetraploidization-related senescence. *International Journal of Molecular Sciences, 24*(7), 6374. https://doi.org/10.3390/ijms24076374.

Certo, M., Moore, V. D. G., Nishino, M., Wei, G., Korsmeyer, S., Armstrong, S. A., & Letai, A. (2006). Mitochondria primed by death signals determine cellular addiction to antiapoptotic BCL-2 family members. *Cancer Cell, 9*(5), 351–365. https://doi.org/10.1016/j.ccr.2006.03.027.

Chang, J., Wang, Y., Shao, L., Laberge, R.-M., Demaria, M., Campisi, J., ... Zhou, D. (2016). Clearance of senescent cells by ABT263 rejuvenates aged hematopoietic stem cells in mice. *Nature Medicine, 22*(1), 78–83. https://doi.org/10.1038/nm.4010.

Chonghaile, T. N., Sarosiek, K. A., Vo, T.-T., Ryan, J. A., Tammareddi, A., Moore, V. D. G., ... Letai, A. (2011). Pretreatment mitochondrial priming correlates with clinical response to cytotoxic chemotherapy. *Science (New York, N. Y.), 334*(6059), 1129–1133. https://doi.org/10.1126/science.1206727.

Di Micco, R., Krizhanovsky, V., Baker, D., & d'Adda di Fagagna, F. (2021). Cellular senescence in ageing: From mechanisms to therapeutic opportunities. *Nature Reviews. Molecular Cell Biology, 22*(2), 75–95. https://doi.org/10.1038/s41580-020-00314-w.

Galiana, I., Lozano-Torres, B., Sancho, M., Alfonso, M., Bernardos, A., Bisbal, V., ... Orzáez, M. (2020). Preclinical antitumor efficacy of senescence-inducing chemotherapy combined with a nanoSenolytic. *Journal of Controlled Release, 323*, 624–634. https://doi.org/10.1016/j.jconrel.2020.04.045.

González-Gualda, E., Baker, A. G., Fruk, L., & Muñoz-Espín, D. (2021). A guide to assessing cellular senescence in vitro and in vivo. *The FEBS Journal, 288*(1), 56–80. https://doi.org/10.1111/febs.15570.

Jia, Y., Han, L., Ramage, C. L., Wang, Z., Weng, C. C., Yang, L., ... Konopleva, M. (2023). Co-targeting BCL-XL and BCL-2 by PROTAC 753B eliminates leukemia cells and enhances efficacy of chemotherapy by targeting senescent cells. *Haematologica, 108*(10), 2626–2638. https://doi.org/10.3324/haematol.2022.281915.

Kale, J., Osterlund, E. J., & Andrews, D. W. (2018). BCL-2 family proteins: Changing partners in the dance towards death. *Cell Death & Differentiation, 25*(1), 65–80. https://doi.org/10.1038/cdd.2017.186.

Kirichenko, T. V., Markina, Y. V., Markin, A. M., Vasilyev, V. S., Hua, H., Li, D., ... Kotenko, K. V. (2025). Functional features of senescent cells and implications for therapy. *International Journal of Molecular Sciences, 26*(11), 5390. https://doi.org/10.3390/ijms26115390.

Kirkland, J. L., & Tchkonia, T. (2020). Senolytic drugs: From discovery to translation. *Journal of Internal Medicine, 288*(5), 518–536. https://doi.org/10.1111/joim.13141.

Maurer, S., Kirsch, V., Ruths, L., Brenner, R. E., & Riegger, J. (2025). Senolytic therapy combining Dasatinib and Quercetin restores the chondrogenic phenotype of human osteoarthritic chondrocytes by the release of pro-anabolic mediators. *Aging Cell, 24*(1), https://doi.org/10.1111/acel.14361.

Montero, J., & Letai, A. (2018). Why do BCL-2 inhibitors work and where should we use them in the clinic? *Cell Death & Differentiation, 25*(1), 56–64. https://doi.org/10.1038/cdd.2017.183.

Ogrodnik, M., Evans, S. A., Fielder, E., Victorelli, S., Kruger, P., Salmonowicz, H., ... Jurk, D. (2021). Whole-body senescent cell clearance alleviates age-related brain inflammation and cognitive impairment in mice. *Aging Cell, 20*(2), https://doi.org/10.1111/acel.13296.

Ryan, J., & Letai, A. (2013). BH3 profiling in whole cells by fluorimeter or FACS. *Methods (San Diego, Calif.), 61*(2), 156–164. https://doi.org/10.1016/j.ymeth.2013.04.006.

Ryan, J., Montero, J., Rocco, J., & Letai, A. (2016). iBH3: Simple, fixable BH3 profiling to determine apoptotic priming in primary tissue by flow cytometry. *Biological Chemistry, 397*(7), 671–678. https://doi.org/10.1515/hsz-2016-0107.

Selt, F., Sigaud, R., Valinciute, G., Sievers, P., Zaman, J., Alcon, C., ... Milde, T. (2023). BH3 mimetics targeting BCL-XL impact the senescent compartment of pilocytic astrocytoma. *Neuro-Oncology, 25*(4), 735–747. https://doi.org/10.1093/neuonc/noac199.

Shahbandi, A., Rao, S. G., Anderson, A. Y., Frey, W. D., Olayiwola, J. O., Ungerleider, N. A., & Jackson, J. G. (2020). BH3 mimetics selectively eliminate chemotherapy-induced senescent cells and improve response in TP53 wild-type breast cancer. *Cell Death & Differentiation, 27*(11), 3097–3116. https://doi.org/10.1038/s41418-020-0564-6.

Singh, R., Letai, A., & Sarosiek, K. (2019). Regulation of apoptosis in health and disease: the balancing act of BCL-2 family proteins. *Nature Reviews. Molecular Cell Biology, 20*(3), 175–193. https://doi.org/10.1038/s41580-018-0089-8.

Wang, B., Han, J., Elisseeff, J. H., & Demaria, M. (2024). The senescence-associated secretory phenotype and its physiological and pathological implications. *Nature Reviews. Molecular Cell Biology, 25*(12), 958–978. https://doi.org/10.1038/s41580-024-00727-x.

Xu, M., Pirtskhalava, T., Farr, J. N., Weigand, B. M., Palmer, A. K., Weivoda, M. M., ... Kirkland, J. L. (2018). Senolytics improve physical function and increase lifespan in old age. *Nature Medicine, 24*(8), 1246–1256. https://doi.org/10.1038/s41591-018-0092-9.

Yamauchi, S., & Takahashi, A. (2025). Cellular senescence: Mechanisms and relevance to cancer and aging. *The Journal of Biochemistry, 177*(3), 163–169. https://doi.org/10.1093/jb/mvae079.

Yang, Y., Jn-Simon, N., He, Y., Sun, C., Zhang, P., Hu, W., ... Zhou, D. (2025). A BCL-xL/BCL-2 PROTAC effectively clears senescent cells in the liver and reduces MASH-driven hepatocellular carcinoma in mice. *Nature Aging, 5*(3), 386–400. https://doi.org/10.1038/s43587-025-00811-7.

Yosef, R., Pilpel, N., Tokarsky-Amiel, R., Biran, A., Ovadya, Y., Cohen, S., ... Krizhanovsky, V. (2016). Directed elimination of senescent cells by inhibition of BCL-W and BCL-XL. *Nature Communications,* 7(1), 11190. https://doi.org/10.1038/ncomms11190.

Zhang, X., Pearsall, V. M., Carver, C. M., Atkinson, E. J., Clarkson, B. D. S., Grund, E. M., ... Schafer, M. J. (2022). Rejuvenation of the aged brain immune cell landscape in mice through p16-positive senescent cell clearance. *Nature Communications, 13*(1), 5671. https://doi.org/10.1038/s41467-022-33226-8.

Zhu, Y., Tchkonia, T., Fuhrmann-Stroissnigg, H., Dai, H. M., Ling, Y. Y., Stout, M. B., ... Kirkland, J. L. (2016). Identification of a novel senolytic agent, navitoclax, targeting the Bcl-2 family of anti-apoptotic factors. *Aging Cell, 15*(3), 428–435. https://doi.org/10.1111/acel.12445.

Zhu, Y., Tchkonia, T., Pirtskhalava, T., Gower, A. C., Ding, H., Giorgadze, N., ... Kirkland, J. L. (2015). The Achilles' heel of senescent cells: from transcriptome to senolytic drugs. *Aging Cell, 14*(4), 644–658. https://doi.org/10.1111/acel.12344.

CHAPTER TWO

Methods to study PIDDosome and caspase-2 signalling

Felix Eichin[a], Rotraud Hirschberger[a], Andreas Villunger[a,b], and Valentina C. Sladky[a,*]
[a]Institute for Developmental Immunology, Biocenter, Medical University of Innsbruck, Innsbruck, Austria
[b]The Research Center for Molecular Medicine (CeMM) of the Austrian Academy of Sciences, Vienna, Austria
*Corresponding author. e-mail address: valentina.sladky@i-med.ac.at

Contents

Abstract

The PIDDosome multiprotein complex is formed by PIDD1 and RAIDD (alias CRADD) and serves as activation platform for caspase-2. One of the best characterized triggers for PIDDosome activation is the presence of extra centrosomes, which are frequently observed in cancer and in naturally polyploid tissues such as liver and heart. Depending on the cell type, activated caspase-2 can (1) cleave MDM2, thereby triggering a p53 response, or (2) cleave the BH3-only protein BID to induce apoptosis. Here, we describe our biochemical and cell biological tools to study the PIDDosome and caspase-2 activity. These include methods to experimentally induce PIDDosome formation in cells, as well as techniques to monitor pathway activity via protein and mRNA expression analysis, and flow cytometry. Moreover, we describe how to follow centrosome maturation for PIDDosome activation using immunofluorescence microscopy.

Methods in Cell Biology, Volume 210
ISSN 0091-679X, https://doi.org/10.1016/bs.mcb.2026.05.012

1. Introduction

The PIDDosome is a multiprotein complex that acts as activation platform for caspase-2. It consists of PIDD1 (P53-Induced Death Domain Protein 1) and RAIDD (RIPK1 domain containing adaptor with death domain, also known as CRADD), which interact via their respective death domains (Tinel & Tschopp, 2004). RAIDD serves as an adapter, as it also contains a CARD (caspase activation and recruitment domain) that enables the interaction with the proform of caspase-2. Upon PIDDosome assembly, pro-caspase-2 is recruited and activated by proximity-induced multimerization and autoproteolytic processing (Fava et al., 2012; Pop & Salvesen, 2009; Tinel & Tschopp, 2004).

Importantly, PIDD1 must also undergo processing to function as a PIDDosome component: Full-length PIDD1 undergoes sequential autoprocessing steps to form the fragments PIDD1-N, PIDD1-C, and PIDD1-CC. The latter autocleavage product, PIDD1-CC, constitutes the core of the PIDDosome responsible for caspase-2 activation (Tinel & Tschopp, 2004; Tinel et al., 2007). Both the PIDDosome complex as well as its individual components have been reported to exert diverse cellular functions, reviewed by us and others elsewhere (Brown-Suedel & Bouchier-Hayes, 2020; Janssens & Tinel, 2012; Sladky et al., 2017; Volik et al., 2023; Weiler et al., 2022). Our research focuses on the role of the PIDDosome as a sensor for polyploidy and centrosome number abnormalities.

The centrosome is a membrane-less cellular organelle that functions as microtubule organising center in most animal cells, forms the poles of the mitotic spindle, and templates ciliogenesis. A diploid cell contains exactly one centrosome that is formed by a pair of centrioles surrounded by pericentriolar material for microtubule nucleation (Breslow & Holland, 2019). To maintain centrosome numbers, centriole duplication occurs exactly once per cell cycle during S-phase, regulated by Polo-like kinase 4 (PLK4). During the subsequent mitosis, the older – mother – centriole acquires distal and subdistal appendages, a number of proteins assembling in a hierarchical manner important for function. Consequently, interphase cells typically carry one centrosome containing an appendage-bearing mother centriole and a newly formed daughter centriole (Nigg & Holland, 2018).

The presence of extra centrosomes as well as whole genome duplication (polyploidy) can promote chromosomal instability and are frequently observed in cancer (Godinho & Basto, 2025; Vittoria et al., 2023). Centrosome amplification (CA), often driven by deregulated PLK4 activity, leads to diploid cells with extra centrosomes. Cell division errors such as aborted mitosis or failed cytokinesis result in whole genome duplication accompanied by centrosome accumulation – a phenomenon commonly observed during tumour evolution. Since the centrosome is duplicated along with DNA, centrosome number serves as a proxy for cellular ploidy (Godinho & Basto, 2025; Nigg & Holland, 2018). The PIDDosome can "count" centrosomes to detect genome-doubled cells (Fava et al., 2017). As such, the PIDDosome regulates proliferation of cells in physiologically polyploid organs such as the liver and the heart and limits the growth of diploid cells with centrosome amplification, as well as that of polyploid cancer cells (Fava et al., 2017; Leone et al., 2026; Sladky et al., 2020).

Specifically, the PIDDosome is activated by the presence of more than one appendage-bearing mother centriole (Fava et al., 2017). Since mother centrioles gain appendages during their first mitosis, PIDDosome activation occurs in the first G1 phase after polyploidization (Fava et al., 2017; Nigg & Holland, 2018). Of note, aberrant increases in PLK4 result in centriole amplification as multiple daughter centrioles are built during S-phase. Thus, it takes one and a half rounds of the cell cycle for these newly formed daughter centrioles to acquire distal appendages (Nigg & Holland, 2018). PIDD1 localizes to the mother centriole via interaction with the distal appendage protein ANKRD26 (Burigotto et al., 2021; Evans et al., 2021). When two or more mother centrioles carrying PIDD1 cluster, the PIDDosome assembles and activates caspase-2, although the mechanistic details and the spatio-temporal dynamics are still unclear (Burigotto et al., 2021). This leads to proteolytic inactivation of the E3-ligase MDM2 resulting in stabilization of the tumour suppressor protein p53. In turn, p53 activates transcription of *CDKN1A/p21* to induce a cell cycle arrest (Fava et al., 2017; Oliver et al., 2011). Alternatively, depending on the cell type, caspase-2 may cleave and activate the pro-apoptotic protein BID to induce apoptotic cell death (Bonzon et al., 2006; Rizzotto et al., 2024).

Here, we present our biochemical and cell biological tools to study the PIDDosome as the central part of the centrosome-p53 axis.

2. Before you begin

Prepare all buffers needed for the respective experiment as detailed in the methods sections below.

3. Materials and equipment

Note: These lists include the general material for all methods presented. Specific materials and reagents, such as antibodies and buffer recipes, are found in the respective methods section. All general material listed here can be substituted with comparable reagents from alternative suppliers.

3.1 Common disposables

- 6 well cell culture plate, flat base (#83.3920.300, Sarstedt)
- 10 cm cell culture dishes (#83.3902, Sarstedt)
- 1.5 mL reaction tubes, polypropylene (#72.690.001, Sarstedt)
- 15 mL screw cap tube, polypropylene (#62.554.016, Sarstedt)
- 50 mL screw cap tube, polypropylene (#62.547.254, Sarstedt)
- High precision cover glass, #1.5H (#0117550, Marienfeld)
- Microscope slide (#HECH42406020, Assistent)

3.2 Cells and reagents

- Cell line of choice, for example, A549 (CCL-185, ATCC), U-2 OS (HTB-96, ATCC), or hTERT RPE-1 (CRL-4000, ATCC)
- Bovine serum albumin, fraction V, low heavy metals (#12659, Sigma-Aldrich)
- Skim milk powder (#70166, Sigma-Aldrich)
- Ponceau S (#141194, Sigma, Aldrich)
- Dulbecco's modified Eagle medium (DMEM), high glucose (#11965092, Thermo Fisher Scientific)
- Fetal bovine serum (#F0804, Sigma-Aldrich)
- L-Glutamine (#G7513, Sigma-Aldrich)
- Penicillin-Streptomycin (#P0781, Sigma-Aldrich)
- Trypsin-EDTA solution (#T4174, Sigma-Aldrich)
- Aurora B kinase inhibitor ZM447439 (#189410, Sigma-Aldrich)
- Dihydrocytochalasin B (DHCB; #D1641, Sigma-Aldrich)
- TRIS-Hydrochloride (#5429.3, Carl Roth GmbH)
- Glycine (#3790.2, Carl Roth GmbH)
- SDS pellets (#CN30.2, Carl Roth GmbH)
- Acrylamide (Rotiphorese®Gel 40, #A515.1 Carl Roth GmbH)
- Temed (N,N,N′,N′-Tetramethyl ethylenediamine, #1.10732, Sigma-Aldrich)
- Ammonium persulfate (APS; #215589, Sigma-Aldrich)
- Tween 20 (#9127.1, Carl Roth GmbH)
- PMSF Protease inhibitor tablet (#P7626, Sigma-Aldrich)

- Amersham Protran Nitrocellulose Membrane (# GE10600002, Sigma-Aldrich)
- Immobilon-P PVDF membrane (#GE10600001, Sigma-Aldrich)
- DNAse I (#DN-25, Sigma-Aldrich)
- Quick-RNA™ MiniPrep (#R1055, Zymo)
- AceQ Universal SYBR qPCR Master Mix (#q511-02, Vazyme)
- Propidium Iodide (#P4170; Sigma)
- 16 % Formaldehyde (w/v), methanol-free (#28908, Thermo Fisher Scientific)
- Methanol for immunofluorescence (#106009, Merck)
- Triton™ X-100 (#1.08603, Sigma-Aldrich)
- Normal Rabbit IgG (#2729, Cell Signalling)
- DAPI (#D1306, Thermo Fisher Scientific)
- Fluoromount (#F4680, Sigma-Aldrich)
- ProLong Gold antifade mountant (#P36930, Invitrogen)

3.3 Equipment

- Bench top centrifuge (Hettich Rotina 420 R)
- Laboratory biological safety cabinet (for example, Herasafe™ KS (NSF) Class II, Type A2, Thermo Fisher Scientific)
- CO_2 Incubator (Heracell™ 150i, Thermo Fisher Scientific)
- Cell counter (CellDrop BF, DeNovix)
- Tweezers
- Aspirator (Ilmvac Biovac 060)
- Mini PROTEAN Tetra Vertical Electrophoresis Cell including gel casting system (#1658001FC, Bio-Rad)
- StepOnePlus RT PCR System (Applied Biosystems)
- Odyssey F Imager (LICORbio)
- Fluorescence microscope
  - For cultured cells: Widefield microscope such as a Thunder Imager (Leica) or ZEISS Axis Observer (Carl Zeiss) with a 63x (1.4) Oil objective.
  - For tissue sections: Confocal microscope, such as the LSM 980 with Airyscan 2 (Carl Zeiss) with a 63x (1.2 NA) glycerol objective.
- Flow cytometer such as Attune NxT flow cytometer (Thermo Fisher Scientific) or LSRFortessa Cell analyzer (BD Biosciences)

3.4 Software

- ImageJ 1.54f (https://imagej.net/ij/)
- Zeiss ZEN or Leica Applications software LAS X

- FlowJo 10
- StepOne Software v2.3
- Image Studio™ Software Version 6.0

4. Step-by-step method details

4.1 Inducing PIDDosome activation

PIDDosome activity is induced by the presence of excess centrosomes, specifically distal appendage-bearing mother centrioles, which can be generated in two ways. First, overexpression of Polo-like kinase 4 (PLK4), the master regulator of centriole biogenesis, results in centrosome amplification in diploid cells (Habedanck et al., 2005; Holland et al., 2012). Second, cells that exit mitosis prematurely or fail cytokinesis become polyploid and carry the corresponding number of centrosomes (Godinho & Basto, 2025). Experimentally, this can be induced by inhibiting Aurora B kinase with ZM447439 for premature abortion of mitosis, or by interfering with the actin cytoskeleton to block cytokinesis using dihydrocytochalasin B (DHCB) as described in the present protocol.

For either induction method, take the doubling time of your cell line of interest into account when selecting time points for analysis. To study PIDDosome activation after PLK4 overexpression, ensure that cells are proliferating (methods described in Garcia-Carpio et al., 2023). PLK4 overexpression drives centriole over-amplification during S-phase, and newly formed daughter centrioles take 1.5 cell cycle rounds to fully mature and acquire distal/subdistal appendages needed for PIDDosome activation (Nigg & Holland, 2018).

Of note, here we focus on the MDM2/p53 axis downstream of the PIDDosome. Caspase-2 mediated cell death via cleavage of the pro-apoptotic protein BID has been described in detail by Rizzotto et al. (Rizzotto et al., 2024). The present protocol precludes experimental details but BaF3 or NALM6 cells are a suitable model and conventional methods for apoptosis detection can be used to study cell death downstream of the centrosome-PIDDosome axis.

4.1.1 Inducing polyploidy

1) Check if your cell line is appropriate for studying PIDDosome function and downstream effects by considering if it has a functional p53 pathway, e.g., not inactivated by the SV40 large T antigen, the E6 protein of Herpes simplex virus, or by mutations. For instance, A549, U2OS, or RPE1 cells are well-suited for studying PIDDosome activation following cytokinesis failure.

2) Seed cells at an appropriate density so that they can proliferate for 2–3 days.
E.g. A549 cells: ~250,000 cells in a 6-well plate. To perform all analyses with the same sample (western blot, flow cytometry, and qPCR), we recommend to seed ~$1.5 * 10^6$ cells in a 10 cm dish.

3) Treat with vehicle control (DMSO) and with ZM447439 (2 μM) or DHCB (4 μM) for 48 h.

4) Harvest the cells with trypsin. Note that for flow cytometry it is critical to have a single cell suspension.
   - **a)** Remove the medium and wash the cells once with PBS. Add 350 μL of trypsin to the 6-well plate and incubate the cells for ~5 min at 37 °C.
   - **b)** Make sure the cells are properly detached from the dish. If not, a longer incubation time is needed to yield a single cell suspension.
   - **c)** Add 1 mL of normal culture medium to stop the digestion with trypsin.
   - **d)** Transfer the cell suspension to a 15 mL tube.

5) Centrifuge the cells at 210 g for 5 min in a pre-cooled centrifuge.

6) Discard the supernatant and wash the cell pellet once with PBS and repeat step 6.

7) Continue with the relevant pathway analyses:
   - **a)** Pathway analysis by Western blot;
   - **b)** Pathway analysis by qRT-PCR;
   - **c)** Pathway analysis by flow cytometer.

4.1.1.1 Troubleshooting and protocol modifications

- DHCB can be quite toxic for certain cell lines. In that case, we recommend using ZM447439 since it is usually well tolerated.

4.2 Monitoring PIDDosome activity

PIDDosome formation results in the activation of caspase-2 by proximity-induced cleavage of pro-caspase-2. Two fragments, p12 and p19, generated from pro-caspase-2 form the active enzyme that then cleaves and thus inactivates MDM2, which causes p53 to accumulate. This, in turn, drives the transcriptional induction of *p21* mRNA and protein (Fava et al., 2017; Oliver et al., 2011; Tinel & Tschopp, 2004). This sequence of events allows us to monitor PIDDosome activity on several levels by immunoblotting: (1) Autoproteolytic activation of caspase-2 can be assessed by the appearance of the cleaved, proteolytically active caspase-2 fragments, p19. However, the p19 fragment is short-lived and thus harder to detect. Alternatively, although an indirect measure, the loss of pro-caspase-2 can also be indicative of PIDDosome-dependent caspase-2 activation. (2) The cleavage products of caspase-2's substrates can be monitored, most prominently cleaved MDM2. (3) The levels of p53 and p21 protein increase after PIDDosome activation (Fava et al., 2017; Oliver et al., 2011; Tinel & Tschopp, 2004). Notably, p53 can be activated by various means (Kastenhuber & Lowe, 2017). Therefore, when monitoring the p53 response as sole readout, it is advisable to always include individual PIDDosome knockout controls, or to treat cells with the specific caspase-2 inhibitor LJ2a (Bosc et al., 2022; Rizzotto et al., 2024). Alternatively, depleting the cells of centrioles using the PLK4 inhibitor centrinone can be used as control condition blocking the pathway upstream of the PIDDosome (Fava et al., 2017; Wong et al., 2015).

In addition to immunoblotting, PIDDosome activation can be monitored by measuring p21 mRNA expression (qRT-PCR), and the subsequent limitation of polyploidization (flow cytometry). The present protocols work for the analysis of cell lines and mouse tissues. Please note that not all listed antibodies (Table 1) work for human and mouse samples.

4.2.1 Materials pathway analysis

Buffers and reagents.

- Random Hexamer (#HP28.1, Carl Roth GmbH)
- MuLV RT buffer (#M0253, New England Biolabs)
- dNTPs (#N0447, New England Biolabs)
- RNAse Inhibitor (#N2511, Promega)
- MuLV reverse transcriptase (#M0253, New England Biolabs)
- **Protein lysis buffer:** 50 mM Tris HCl pH 8.0, 150 mM NaCl, 0.5 % NP-40, 50 mM NaF, 1 mM Na_3VO_4, 1 mM PMSF, 1 tablet Protease inhibitor, 10 mg/mL DNase I, add deionized H_2O (dH_2O) up to 10 mL
- **5x Laemmli sample buffer:** 250 mM TRIS pH 6.8, 10 % SDS, 0.5 % bromophenol blue, 50 % Glycerol, add 500 mM 2-Mercaptoethanol before use
- **Separation Buffer:** 1 M TRIS/HCl, pH 8.8
- **Stacking Buffer:** 0.375 M TRIS/HCl, pH 6.8

Table 1 Antibodies to investigate the PIDDosome pathway by Western blot. Dilutions should be made in PBS, containing 5 % BSA, 0.1 % sodium azide and 0.1 % Tween.

Primary Antibodies

Target	*Provider*	*Catalog #*	*Clone*	*Species*	*Reactivity*	*Dilution*
Caspase-2	Enzo Life Science	ALX-804-356	11B4	rat	mouse, human	1:500
HSP90	Santa Cruz	sc-13119	F8	mouse	mouse, human	1:1000
MDM2	Invitrogen	33-7100	IF2	mouse	human	1:1000
MDM2	Cell Signaling	51541	E3G5I	rabbit	mouse, human	1:1000
P21	Cell Signaling	2947	12D1	rabbit	mouse, human	1:1000
P53	Cell Signaling	2524	1C12	mouse	mouse, human	1:1000
PIDD1	Enzo Life Science	ALX-804-837	Anto-1	mouse	human	1:1000
PIDD1	ProSci	36-150	Lise-1	mouse	mouse	1:1000
RAIDD	Proteintech	10401-1-AP	poly	rabbit	mouse, human	1:1000

Secondary Antibodies

Target	*Provider*	*Catalog #*	*conju-gated*	*Species*		*Dilution*
anti-mouse	Agilent,	P026002-2	Hrp	goat		1:10,000
anti-rabbit	Agilent	P044801-2	Hrp	goat		1:10,000
anti-rat	Cell Signaling	7077S	Hrp	goat		1:10,000
anti-mouse IgG light chain	Cell Signaling	58802S	Hrp	rabbit		1:10,000

- **10x PAGE running buffer:** 26 g TRIS, 144 g Glycine, 10 g SDS adjusted to 1 L with dH_2O
- **10 % separation gel:** 2.5 mL Acrylamide (40 %), 3.76 mL separation buffer, 3.52 mL dH_2O, 100 µL 10 % SDS, 125 µL 10 % APS, 2.5 µL Temed
- **12 % separation gel:** 3.0 mL Acrylamide (40 %), 3.76 mL separation buffer, 3.02 mL dH_2O, 100 µL 10 % SDS, 125 µL 10 % APS, 2.5 µL Temed
- **Stacking gel:** 0.625 mL Acrylamide (40 %), 2.1 mL stacking buffer, 3.0 mL dH_2O, 62.5 µL 10 % SDS, 250 µL 10 % APS, 2.5 µL Temed
- **Transfer buffer:** 3.03 g TRIS, 14.4 g Glycine, 200 mL EtOH, adjust to 1 L with dH_2O
- **Western blot blocking buffer:** 5 % skim milk powder in 0.1 % PBS-Tween
- **Western blot washing buffer:** 0.1 % Tween in PBS

Antibodies and oligonucleotides.

4.2.2 Pathway analysis by Western blot

Timing: Day 1 – protein extraction and gel preparation: 3–4 h, Day 2 – SDS-PAGE and Western blotting: 4–4.5 h, Day 3 – Detection: 1 h Table 2.

1. Protein extraction with protein lysis buffer.
   Note: For tissue samples we recommend conventional RIPA lysis buffer.
   a. Re-suspend the cell pellet in lysis buffer and incubate at least 30 min on ice.
   b. Centrifuge at full speed for 10 min in a pre-cooled (4 °C) centrifuge.
   c. Transfer supernatant in a new 1.5 mL reaction tube.
   d. Determine protein concentration (e.g., using Bradford Assay).
   e. Adjust the samples to the same protein concentration and add Laemmli Buffer to a final concentration of 1x.
2. Prepare the appropriate polyacrylamide gels.
   Note: For p21 and RAIDD detection, a 12 % gel should be used!

Table 2 Primer sequences for qRT-PCR.

Target	**Sequence**
human p21	5′AGTCAGTTCCTTGTGGAGCC 3′ 5′CATTAGCGCATCACAGTCGC 3′
human Hprt	5′ GCTATAAATTCTTTGCTGACCTGCTG 3′ 5′ AATTACTTTTATGTCCCCTGTTGACTGG 3′

3. Load equal amounts of protein (best 40–60 μg) on the gel and run it with constant 100–130 V until the front is running out.
   Note: pay attention that small proteins like RAIDD and p21 are not running out of the gel.
4. Blot the proteins via a wet transfer onto a nitrocellulose membrane. Running conditions: constant 100 V for 80 min. Cool the transfer chamber.
   Note: Nitrocellulose membranes work well for the detection of caspase-2, RAIDD, MDM2, p53, and p21. For detecting PIDD1, a PVDF membrane can work better, but our experiences are mixed.
5. Check transfer quality by Ponceau S staining
   a. Stain the membrane with Ponceau S solution for 1 min
   b. Wash membrane 2–3 times with dH_2O.
   c. De-stain the membrane with washing buffer.
6. Block the membrane with blocking buffer for 1 h at room temperature.
7. Add the primary antibody (Table 1) and incubate at 4 °C overnight.
   Note: If you want to detect proteins of similar size on the same membrane, start with the one with the weakest expected signal. E.g., start with PIDD1.
8. Collect primary antibody for reuse and wash the membrane 5 times for 5 min each with washing buffer at room temperature.
   Note: The primary antibody dilutions can be reused until the signal intensity starts to decrease. The caspase-2 and the HSP90 antibodies can only be reused 2–3 times, as the signal strength decreases fast. All other mentioned antibodies can be reused at least 5 times, depending on the number of samples and signal intensities detected.
9. Incubate the membrane with the respective 2nd antibody diluted in blocking buffer at room temperature for 1 h.
10. Discard the 2nd antibody solution and wash again 5 times for 5 min
11. Detect the signal of the secondary antibody.
12. Repeat step 9–11 for all targets/antibodies.

4.2.2.1 Troubleshooting and protocol modifications

- The loading control (HSP90), RAIDD, and – depending on the cell line or tissue – pro-caspase-2 usually yield a strong signal. Detecting PIDD1 as well as the cleaved, active fragment of caspase-2 (p19) is more difficult. For the latter we recommend the most sensitive detection method available. Using X-ray films with the respective HRP-conjugated secondary antibodies and enhanced chemiluminescence solution (ECL) is a good alternative if the Licor Odyssey system is not available. In our hands, other chemiluminescence detection systems were not sensitive enough to pick up the active caspase-2 fragment.

- Pro-caspase-2 is autoproteolytically cleaved into the p32 fragment and the p19/p12 fragments. The latter ones correspond to the active form of caspase-2 and are very short-lived and therefore hard to detect. Importantly, the p32 intermediate fragment has been found to be an unreliable measure for caspase-2 activation. In our hands, together with detecting MDM2 cleavage or p53 stabilisation, loss of pro-caspase-2 is an indirect but robust readout for caspase-2 activation in this context.
- For detecting the short-lived p19 caspase-2 fragment, we recommend performing a time course experiment as the peak of caspase-2 activity varies between cell lines and activation triggers. Moreover, we recommend using the most sensitive detection method at hand.
- MDM2 is a transcriptional target of p53, so the overall signal for MDM2 increases whenever p53 is activated. Reflecting this positive feedback loop, in this assay the bands corresponding to the MDM2 cleavage fragments are often much stronger than the full-length MDM2 in the untreated sample.
- Depending on the cell line or tissue, the MDM2 antibody picks up unspecific background bands. To determine which bands correspond to the caspase-2 generated cleavage products, it is critical to include a DMSO and a ZM447439 (or DHCB) treated wildtype control cell line sample.
- If mouse tissue samples are analysed, endogenous mouse antibodies can be present in the lysate in varying abundance depending on the tissue. When using primary antibodies raised in mouse on such samples, the anti-mouse secondary antibody recognizes also the endogenous IgG light chain and heavy chain. This results in a smear-like signal on the blot that can then mask the band of interest since the IgG heavy chain runs around 55 kDa – a size at which both caspase-2 and p53 are found. For monitoring PIDDosome activity, this is specifically relevant when detecting p53 using the recommended murine primary antibody. To overcome this issue, a light-chain specific secondary antibody can be used for targets of a size about 55 kDa.

4.2.3 Pathway analysis by qRT-PCR

Timing: 3–3.5 h

1. Extract RNA from the cells using the RNA extraction kit.
2. Determine RNA concentration and quality using a photometer.

3. cDNA synthesis:
   a. Mix 100 ng of RNA with 125 ng random Hexamer (#HP28.1, Carl Roth GmbH) in 5 μL dH_2O and incubate the mix for 5 min at 65 °C.
   b. Mix with 5 μL of reverse transcriptase mix (1 μL MuLV RT buffer #M0253 NEB, 0.5 μL dNTPs (10 mM each, #N0447 NEB), 0.25 μL RNAse Inhibitor #N2511 Promega, 0.25 μL MuLV reverse transcriptase #M0253 NEB, 3 μL H_2O) and incubate for 5 min at 25 °C and then for 1 h at 42 °C.
   c. Add 90 μL H_2O and use 2 μL of the cDNA mix for one qPCR reaction.
4. Perform the qRT-PCR with duplicates or triplicates for each sample and target using 2 μL of your cDNA.

4.2.3.1 Troubleshooting and protocol modifications

- In addition to the dH_2O control, include a negative control for each target, e.g. cDNA from knockout cells. To minimize errors caused by pipetting, we recommended to prepare master mixes for each target and cDNA sample.
- We use SYBR (Vazyme) for qPCRs since it is less costly but probe-based assays work equally well.
- Expression analysis of p21 reflects general p53 activity. Thus, monitoring of the PIDDosome pathway by qRT-PCR should be combined with additional readouts such as immunoblotting for MDM2 cleavage products. This is exemplified in Fig. 1B: p21 expression is also induced to a certain degree in *Pidd1*$^{-/-}$ cells upon cytokinesis failure, likely due to mitotic errors and/or DNA damage caused by ZM447439 (Sladky et al., 2020).

4.2.4 Pathway analysis by flow cytometry

Timing: Day 1 – harvest and fixation: 1 h, Day 2 – staining and analysis: 2–4 h, depending on the number of samples.

1. Re-suspend the cells in 300 μL PBS and transfer them into a 5 mL tube.
2. Add 700 μL of ice-cold 100 % ethanol while vortexing to the cell suspension (final EtOH conc. 70 %).
3. Store the cell-ethanol suspension at least overnight at −20 °C.
   Note: The cells can be stored for several days/weeks if necessary. If stored for longer than a few days, the tubes should be sealed to prevent drying.
4. Add 2 mL PBS to the cell-ethanol suspension and centrifuge at 470 g for 5 min

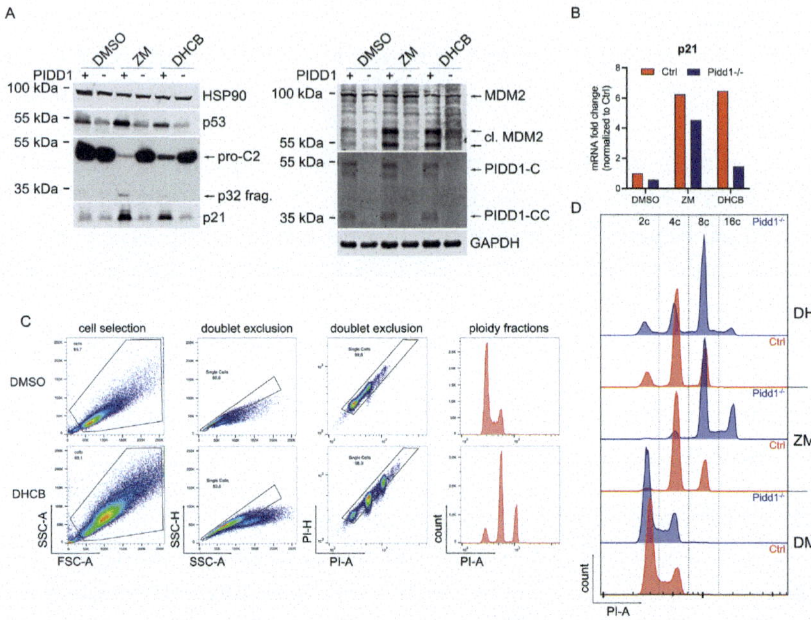

Fig. 1 Analysis of the PIDDosome pathway. (A) Western blot analysis of parental (PIDD1 +) or PIDD1-deficient (PIDD1-, CRISPR/Cas9 mediated knockout) A549 cells treated with 2 μM ZM447439 or 4 μM dihydrocytochalasin B (DHCB) for 48 h. (B) Expression analysis of p21 mRNA of the samples from A via qRT-PCR. (C-D) Gating strategy and ploidy analysis of the A549 cells from A with propidium iodide staining and flow cytometry. **C** Representative gating of control cells (CRISPR control). (D) Histogram overlays of the cells treated with the indicated drugs for 48 h.

5. Discard the supernatant by inverting the tube and washing the cells again with 2 mL PBS.
6. Re-suspend the cells in 300 μL PBS propidium iodide (PI) solution (33.3 μg/mL PI and 0.01 μg/mL RNase A) and incubate for 15 min at 37 °C.
   Note: Instead of PI, other DNA dyes such as DAPI can be used.
7. Pass the sample through a 200 μm filter to prevent clogging of the flow cytometer.
8. Analyse the DNA content on a flow cytometer.
   Note: Make sure to record the height, width and area parameters of the needed channels. This is required for proper doublet exclusion!
   Note: Acquire in log scale to record also highly polyploid cells.
9. Gating strategy (as outlined in Fig. 1C):
   a. FSC-A vs SSC-A – gate for cells.

b. First doublet exclusion: SSC-H vs SSC-A – single cells are found on a straight line diagonally over the plot.
c. Second doublet exclusion based on the PI staining for DNA: PI-H vs PI-A – single cells form individual ploidy populations with S-phase cells in between. Dots found below the diagonal represent doublets.
d. Ploidy fractions: histogram for PI-H or PI-A – the peaks represent the different ploidy populations and can be gated to measure the percentage.

4.2.4.1 Troubleshooting and protocol modifications

- Always include an untreated or DMSO-treated control sample as reference to correctly identify the 2 n and 4 n peaks. For your untreated sample, the first sharp peak corresponds to 2 n/G1 cells and the second one to 4 n/G2/M cells. The intensities of the peaks should double, as does DNA with every round of failed cytokinesis (2 n, 4 n, 8 n, etc.). Smaller peaks in between can be caused by doublets and disappear with a stringent doublet exclusion. Highly aneuploid cells or cells in S phase are also found between the major peaks. Dead cells appear as broad 'sub-G1' peak. Note that this is not a suitable method to quantify aneuploidy.
- PIDDosome pathway-deficient cells can re-enter the cell cycle after cytokinesis failure and fail again, thus becoming highly polyploid (Fig. 1D). In contrast, PIDDosome-proficient cells reduce proliferation after the first round of failed cytokinesis, i.e. in a tetraploid state. However, the efficiency of this cell cycle arrest varies between cell lines. For example, A549 cells largely arrest in a tetraploid state, wildtype MEFs or RPE-1 cells show reduced proliferation and polyploidization compared to the respective PIDDosome-deficient cells.
- Make sure to properly dissociate adherent cells or cells isolated from mouse tissues to generate a true single-cell suspension. Cell clumps can clog the flow cytometer and make the ploidy analysis unreliable.
- The DNA staining can be combined with immunostaining for the target of interest. However, we recommend fixing the cells using ethanol or methanol, as paraformaldehyde fixation can promote cell clumping blurring the DNA profile.

4.2.5 Expected outcomes

For PIDDosome-proficient cells, PIDDosome activation should be detectable following failed cell division due to ZM447439 or DHCB treatment using the following methods. (1) Western blotting: appearance of bands corresponding to cleaved MDM2, loss of pro-caspase-2 and appearance of the p19 fragment (for the latter, see troubleshooting advice), p53, and p21. (2) qRT-PCR: upregulation of the p21 mRNA compared to the DMSO control. (3) Flow cytometry: a shift in ploidy with most cells in the tetraploid fraction. The exact ploidy distribution depends on the cell line and the timepoint of analysis.

For PIDDosome-deficient cells, PIDDosome-dependent signaling is expected to be fully abrogated. Therefore, flow cytometric analysis should reveal a massive ploidy increase following ZM447439 or DHCB treatment. Examples of all analyses and outcomes are found in Fig. 1.

4.3 Monitoring the upstream signal – visualizing ANKRD26, PIDD1, and the centrosome by immunofluorescence

Imaging of centrioles, distal appendage proteins and PIDD1 can elucidate if cells fulfil the prerequisites for PIDDosome activation: (1) the presence of at least two appendage-carrying mother centrioles and (2) clustering of mother centrioles (Burigotto et al., 2021; Evans et al., 2021). To this end, we use immunofluorescence microscopy to visualize all centrioles, as well as distal appendages where PIDD1 localizes. The centriole distal appendage is a tree-like structure formed by the proteins C2CD3, CEP83, and SCLT1. The latter recruits FBF1, CEP164 and ANKRD26, which binds PIDD1 (Bowler et al., 2019; Breslow & Holland, 2019; Burigotto et al., 2021; Evans et al., 2021).

We also stain for the centriole capping protein CP110 to detect all centrioles. To determine the number of distal appendage-bearing centrioles, we recommend using antibodies targeting ANKRD26 or CEP164. Of note, mother centrioles acquire appendages during mitotic traverse (Nigg & Holland, 2018). Hence, about 50 % of centrioles are expected to be positive for ANKRD26 and CEP164. Assessing the ratio all centrioles:mother centrioles can serve as an internal reference for the staining quality. This is especially important when performing the tertiary staining protocol with antibodies raised in the same species.

The presence of ANKRD26 and PIDD1 on centrioles and the number of those centrioles can inform directly on the ability of cells to activate the PIDDosome (Burigotto et al., 2021; Evans et al., 2021). Unfortunately, no reliable antibodies for detection of mouse ANKRD26 or mouse PIDD1 by immunofluorescence are currently available. For murine cell lines and tissues – although an indirect measure – CEP164 staining can be used instead to assess the number of mother centrioles (Graser et al., 2007).

The following protocols work for cells on coverslips as well as for unfixed tissue cryosections. For fixed tissue cryosections, an appropriate antigen retrieval protocol may be needed (such as found in Phan et al., 2021). Note that not all of the recommended antibodies work for murine proteins (see Table 3).

4.3.1 Materials immunofluorescence

Buffers and reagents.

- **Blocking buffer:** 2.5 % FBS, 200 mM Glycine, and 0.1 % Triton X-100 in PBS
- **Washing buffer:** PBST – 0.1 % Triton X-100 in PBS
- **PTEM buffer:** 0.2 % Triton™ X-100, 20 mM PIPES at pH 6.8, 1 mM MgCl2, 10 mM EGTA in ddH_2O
- **DAPI working solution:** 1 mg/mL DAPI in PBS, use 1:1000

Antibodies.

Table 3 Antibodies used for immunofluorescence staining.

Primary Antibodies

Target	*Provider*	*Catalog #*	*Clone*	*Species*	*Reactivity*	*Dilution*
CP110	Proteintech	12780-1-AP	polyclonal	rabbit	mouse, human	1:500
γTubulin	Santa Cruz	sc−51715	T30	mouse	mouse, human	1:250
ANKRD26	Genetex	GTX128255	polyclonal	rabbit	human	1:1000
PIDD1	Enzo	ALX-804-837	Anto-1	mouse	human	1:500
CEP164	Merck Millipore	ABE2126	polyclonal	rabbit	mouse, human	1:1000

Secondary Antibodies

Target	*Provider*	*Catalog #*	*conjugate*	*Species*	*Dilution*
anti-mouse IgG	Thermo Fisher Scientific	A-21236	Alexa Fluor™ 647	goat	1:5000
anti-rabbit IgG	Thermo Fisher Scientific	A-11034	Alexa Fluor™ 488	goat	1:5000

4.3.2 Standard immunofluorescence staining

Timing: 4.5–5 h excluding image acquisition time.

1. Wash cells or tissue with PBS.
2. Pre-extract with PTEM buffer for 2 min at RT.
3. Fix with 4 % formaldehyde in PBS for 10 min at RT.
4. Rinse twice with PBS.

   Optional: Quench with 50 mM NH_3Cl for 15 min at RT (recommended for cells with high autofluorescence).
5. Block in blocking buffer for 1 h at RT or longer at 4 °C.
6. Stain with primary antibody in blocking buffer 1 h at RT.
7. Wash 3x with PBST for 5 min each time.
8. Stain with secondary antibody in blocking buffer containing DAPI (1 μg/mL final concentration) 1 h at RT (in wells).
9. Wash 3x with PBST for 5 min each time.
10. Rinse with PBS.
11. Mount with Fluoromount G or Prolong Gold.
12. Image coverslips on a fluorescence microscope equipped with a 63x or 100x objective.

Note: Since centrioles are small structures (~250 × 500 nm), depending on the microscope it can be beneficial to acquire Z-stacks. For confocal imaging, use the optimal step size for the respective objective, at least 0.3 μm.

4.3.3 Tertiary immunofluorescence staining

This protocol can be used to stain with two different primary antibodies that were raised in the same species, e.g. *the antibodies targeting CP110 and ANKRD26, both raised in rabbit. Treatment with rabbit IgG after the first round of staining blocks the epitopes of the first rabbit antibody added. Thus, in the second round of staining, the secondary anti-rabbit antibody will only bind the second primary antibody.*

Timing: 7.5–8 h excluding image acquisition time.

1. Wash cells or tissue with PBS.
2. Pre-extract with PTEM buffer for 2 min at RT.
3. Fix with 4 % formaldehyde in PBS for 10 min at RT.
4. Rinse twice with PBS.

   Optional: Quench with 50 mM NH_3Cl for 15 min at RT (recommended for cells with high autofluorescence).
5. Block in blocking buffer for 1 h at RT or longer at 4 °C.
6. Stain with first primary antibody in blocking buffer 1 h at RT.

7. Wash 3x with PBST for 5 min each time.
   Note: For tissue cryosections, increase the Triton X concentration in the washing buffer to 0.5%.
8. Stain with first secondary antibody in blocking buffer 1 h at RT.
9. Wash 3x with PBST for 5 min each time.
10. Incubate with Rabbit IgG 30 min at RT.
11. Wash 3x with PBST for 5 min each time.
12. Stain with second primary antibody in blocking buffer 1 h at RT.
13. Wash 3x with PBST for 5 min each time.
14. Stain with second secondary antibody and DAPI (1 μg/mL final concentration) in blocking buffer 1 h at RT.
15. Wash 3x with PBST for 5 min each time.
16. Rinse with PBS.
17. Mount with Fluoromount or Prolong Gold.
18. Image as described above.

4.3.3.1 Troubleshooting and protocol modifications

- For tissue cryosections, increase the Triton X concentration in the washing buffer to 0.5 %.
- For cells with high autofluorescence, include the optional quenching step.
- Depending on the cell type or tissue, an alternative fixation method can improve the results and reduce background. Replace steps 3−4 with the following fixation protocol:
  1. Fix for 4 min in 1.5 % formaldehyde in PBS.
  2. Aspirate the formaldehyde solution.
  3. Directly add ice-cold 100 % Methanol for 4 min at −20 °C.
  4. Rinse with PBS twice.
- As an internal reference, check whether approximately half of all CP110 foci (i.e., all centrioles) are positive for ANKRD26 or CEP164. This is especially important for the tertiary staining. Overlapping signals for CP110 and ANKRD26/CEP164 on all centrioles indicate that blocking with rabbit IgG was insufficient.

4.3.4 Expected outcomes

The resulting immunofluorescence images show cells with all centrioles (CP110) as well as the target of interest, i.e. ANKRD26 or PIDD1, similar to the examples shown in Fig. 2. These proteins are found on the distal appendage of the mother centrioles. Thus, in the control treatment most

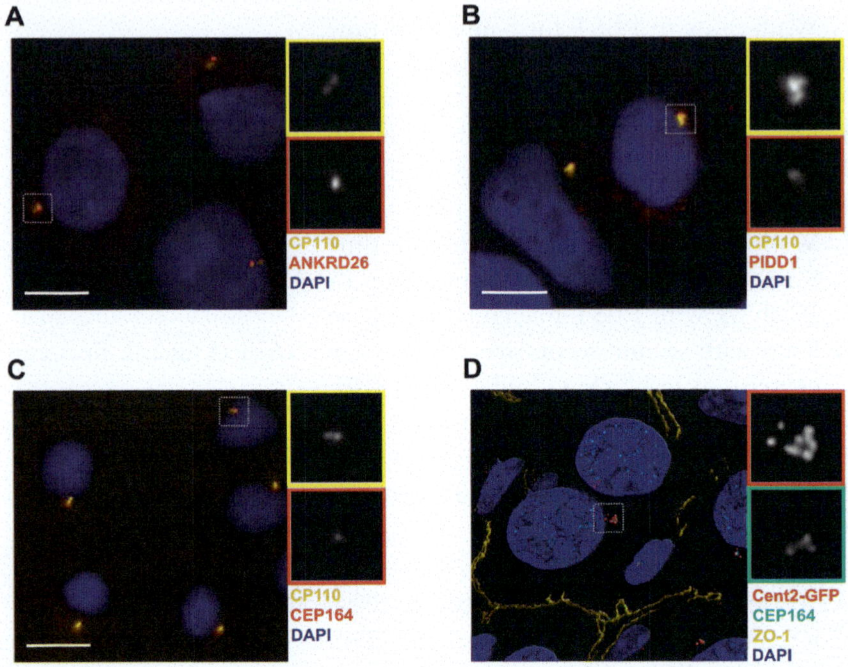

Fig. 2 Examples of immunofluorescence microscopy to visualize centrosomal proteins. (A-C) A549 cells were immuno-stained with CP110 to mark all centrioles and with antibodies recognizing (A) ANKRD26, (B) PIDD1, and (C) CEP164 to visualize distal appendages. Scale bars 10 μm. (D) Micrograph showing a mouse liver cryosection stained with antibodies targeting CEP164 and ZO-1 to stain cell boundaries. All centrioles are positive for transgenic Centrin-2-GFP.

cells should harbor one focus of ANKRD26, PIDD1, or CEP164. Cells that have undergone cell division failure after ZM447439 or DHCB treatment should carry at least two foci corresponding to those proteins. In PIDDosome-deficient treated cells, which are usually highly polyploid, a higher number of centrioles (CP110) and mother centrioles can be expected.

5. Quantification and statistical analysis

We recommend performing at least an n = 3 biological replicates for each condition for all methods presented. For the pathway analysis by western blotting, the resulting bands can be quantified densitometrically. For qRT-PCR, we us the $2^{-\Delta\Delta Ct}$ method to determine fold induction relative to the DMSO control. Flow cytometry data are analysed according to the described gating strategy. Apply the same gates to all biological replicates and quantify the different ploidy fractions as shown in Fig. 1D. To quantify immunofluorescence images, per cell count all centrioles (stained by CP110) and the number of foci corresponding to the target of interest such as ANKRD26 or PIDD1.

6. Conclusion

Since the discovery of the PIDDosome and its role in activating caspase-2 in 2004, research on the PIDDosome has advanced significantly, particularly in recent years (Brown-Suedel & Bouchier-Hayes, 2020; Sladky et al., 2017; Tinel & Tschopp, 2004; Volik et al., 2023). Activated by extra centrosomes, the PIDDosome plays a critical role in (1) blocking the proliferation of polyploid, and thus potentially chromosomally unstable, cells, and (2) regulating ploidy during organogenesis and tissue homeostasis (Fava et al., 2017; Rizzotto et al., 2024; Sladky et al., 2020).

However, future research requires the development and refinement of materials and tools to study the PIDDosome in various biological contexts. Although upstream triggers for PIDDosome activation – most prominently excess mature centrioles – have been described, the spatiotemporal dynamics of complex assembly remain poorly understood. Efforts to visualize caspase-2 and RAIDD in living cells have been unsuccessful, highlighting the need for alternative tools.

Another key area requiring further investigation is the identification and validation of caspase-2 substrates. While the caspase-2-mediated cleavage of MDM2 and BID has been well characterized and confirmed to be physiologically relevant, many other proposed substrates await in-depth validation in both in vitro and in vivo to determine their processing context, as well as caspase-2 selectivity. Unbiased proteomics approaches may aid in identifying novel substrates relevant to specific tissues, developmental stages, or diseases.

A promising novel tool is a recently developed selective inhibitor for caspase-2 (Bosc et al., 2022). Alongside this, the generation of small-molecule inhibitors and activators for caspase-2 and the PIDDosome will further promote in vivo research and may pave the way for future therapeutic applications.

Acknowledgements

We thank all members of the Institute of Developmental Immunology as well as Prof. Luca Fava (CIBIO, Trento, IT) and Dr. Thao P. Phan (UCSF, USA) for discussions and protocol optimization over the years. This research was funded in part by the Austrian Science Fund (FWF) to AV [10.55776/DOC82] and to VCS [10.55776/ESP1232724].

References

Bonzon, C., Bouchier-Hayes, L., Pagliari, L. J., Green, D. R., & Newmeyer, D. D. (2006). Caspase-2–induced apoptosis requires bid cleavage: A physiological role for bid in heat shock–induced death. *Molecular Biology of the Cell, 17*, 2150–2157. https://doi.org/10.1091/mbc.e05-12-1107.

Bosc, E., Anastasie, J., Soualmia, F., Coric, P., Kim, J. Y., Wang, L. Q., et al. (2022). Genuine selective caspase-2 inhibition with new irreversible small peptidomimetics. *Cell Death & Disease, 13*, 959. https://doi.org/10.1038/s41419-022-05396-2.

Bowler, M., Kong, D., Sun, S., Nanjundappa, R., Evans, L., Farmer, V., et al. (2019). High-resolution characterization of centriole distal appendage morphology and dynamics by correlative STORM and electron microscopy. *Nature Communications, 10*(1), 1–15. https://doi.org/10.1038/s41467-018-08216-4.

Breslow, D. K., & Holland, A. J. (2019). Mechanism and regulation of centriole and cilium biogenesis. *Annual Review of Biochemistry, 88*, annurev-111153. https://doi.org/10.1146/annurev-biochem-013118-111153.

Brown-Suedel, A. N., & Bouchier-Hayes, L. (2020). Caspase-2 substrates: To apoptosis, cell cycle control, and beyond. *Frontiers in Cell and Developmental Biology, 8*. https://doi.org/10.3389/fcell.2020.610022.

Burigotto, M., Mattivi, A., Migliorati, D., Magnani, G., Valentini, C., Roccuzzo, M., et al. (2021). Centriolar distal appendages activate the centrosome-PIDDosome-p53 signalling axis via ANKRD26. *The EMBO Journal*, 1–22. https://doi.org/10.15252/embj.2020104844.

Evans, L. T., Anglen, T., Scott, P., Lukasik, K., Loncarek, J., & Holland, A. J. (2021). ANKRD26 recruits PIDD1 to centriolar distal appendages to activate the PIDDosome following centrosome amplification. *The EMBO Journal*, 1–18. https://doi.org/10.15252/embj.2020105106.

Fava, L. L., Bock, F. J., Geley, S., & Villunger, A. (2012). Caspase-2 at a glance. *Journal of Cell Science, 125*, 5911–5915. https://doi.org/10.1242/jcs.115105.

Fava, L. L., Schuler, F., Sladky, V., Haschka, M. D., Soratroi, C., Eiterer, L., et al. (2017). The PIDDosome activates p53 in response to supernumerary centrosomes. *Genes and Development, 31*, 34–45. https://doi.org/10.1101/gad.289728.116.

Garcia-Carpio, I., Braun, V. Z., Weiler, E. S., Leone, M., Niñerola, S., Barco, A., et al. (2023). Extra centrosomes induce PIDD1-mediated inflammation and immunosurveillance. *The EMBO Journal, 42*, e113510. https://doi.org/10.15252/embj.2023113510.

Godinho, S. A., & Basto, R. (2025). Centrosomes and cancer: Balancing tumor-promoting and inhibitory roles. *Trends in Cell Biology, 35*, 515–526. https://doi.org/10.1016/j.tcb.2025.02.009.

Graser, S., Stierhof, Y. D., Lavoie, S. B., Gassner, O. S., Lamla, S., Le Clech, M., et al. (2007). Cep164, a novel centriole appendage protein required for primary cilium formation. *Journal of Cell Biology, 179*, 321–330. https://doi.org/10.1083/jcb.200707181.

Habedanck, R., Stierhof, Y.-D., Wilkinson, C. J., & Nigg, E. A. (2005). The Polo kinase Plk4 functions in centriole duplication. *Nature Cell Biology, 7*, 1140–1146. https://doi.org/10.1038/ncb1320.

Holland, A. J., Fachinetti, D., Zhu, Q., Bauer, M., Verma, I. M., Nigg, E. A., et al. (2012). The autoregulated instability of Polo-like kinase 4 limits centrosome duplication to once per cell cycle. *Genes & Development, 26*, 2684–2689. https://doi.org/10.1101/gad.207027.112.

Janssens, S., & Tinel, A. (2012). The PIDDosome, DNA-damage-induced apoptosis and beyond. *Cell Death & Differentiation, 19*, 13–20. https://doi.org/10.1038/cdd.2011.162.

Kastenhuber, E. R., & Lowe, S. W. (2017). Putting p53 in Context. *Cell, 170*, 1062–1078. https://doi.org/10.1016/J.CELL.2017.08.028.

Leone, M., Kinz, N., Eichin, F., Obwegs, D., Sladky, V. C., Braun, V. Z., et al. (2026). The PIDDosome controls cardiomyocyte polyploidization during postnatal heart development. *Cell Death & Differentiation*. https://doi.org/10.1038/s41418-025-01645-x.

Nigg, E. A., & Holland, A. J. (2018). Once and only once: mechanisms of centriole duplication and their deregulation in disease. *Nature Reviews. Molecular Cell Biology, 19*, 297–312. https://doi.org/10.1038/nrm.2017.127.

Oliver, T. G., Meylan, E., Chang, G. P., Xue, W., Burke, J. R., Humpton, T. J., et al. (2011). Caspase-2-Mediated Cleavage of Mdm2 Creates a p53-Induced Positive Feedback Loop. *Molecular Cell, 43*, 57–71. https://doi.org/10.1016/j.molcel.2011.06.012.

Phan, T. P., Maryniak, A. L., Boatwright, C. A., Lee, J., Atkins, A., Tijhuis, A., et al. (2021). Centrosome defects cause microcephaly by activating the 53BP1-USP28-TP53 mitotic surveillance pathway. *The EMBO Journal, 40*, e106118. https://doi.org/10.15252/embj.2020106118.

Pop, C., & Salvesen, G. S. (2009). Human Caspases: Activation, Specificity, and Regulation. *Journal of Biological Chemistry, 284*, 21777–21781. https://doi.org/10.1074/JBC.R800084200.

Rizzotto, D., Vigorito, V., Rieder, P., Gallob, F., Moretta, G. M., Soratroi, C., et al. (2024). Caspase-2 kills cells with extra centrosomes. *Science Advances, 10*, eado6607. https://doi.org/10.1126/sciadv.ado6607.

Sladky, V., Schuler, F., Fava, L. L., & Villunger, A. (2017). The resurrection of the PIDDosome – emerging roles in the DNA-damage response and centrosome surveillance. *Journal of Cell Science, 130*(22), 37. https://doi.org/10.1242/jcs.203448.

Sladky, V. C., Knapp, K., Soratroi, C., Heppke, J., Eichin, F., Rocamora-Reverte, L., et al. (2020). E2F-Family Members Engage the PIDDosome to Limit Hepatocyte Ploidy in Liver Development and Regeneration. *Developmental Cell, 52*, 335–349.e7. https://doi.org/10.1016/j.devcel.2019.12.016.

Tinel, A., Janssens, S., Lippens, S., Cuenin, S., Logette, E., Jaccard, B., et al. (2007). Autoproteolysis of PIDD marks the bifurcation between pro-death caspase-2 and pro-survival NF-κB pathway. *EMBO Journal, 26*, 197–208. https://doi.org/10.1038/sj.emboj.7601473.

Tinel, A., & Tschopp, J. (2004). The PIDDosome, a protein complex implicated in activation of caspase-2 in response to genotoxic stress. *Science (New York, NY), 304*, 843–846. https://doi.org/10.1126/science.1095432.

Vittoria, M. A., Quinton, R. J., & Ganem, N. J. (2023). Whole-genome doubling in tissues and tumors. *Trends in Genetics, 0*. https://doi.org/10.1016/j.tig.2023.08.004.

Volik, P. I., Kopeina, G. S., Zhivotovsky, B., & Zamaraev, A. V. (2023). Total recall: the role of PIDDosome components in neurodegeneration. *Trends in Molecular Medicine, 29*, 996–1013. https://doi.org/10.1016/j.molmed.2023.08.008.

Weiler, E. S., Szabo, T. G., Garcia-Carpio, I., & Villunger, A. (2022). PIDD1 in cell cycle control, sterile inflammation and cell death. *Biochemical Society Transactions, 50*, 813–824. https://doi.org/10.1042/BST20211186.

Wong, Y. L., Anzola, J. V., Davis, R. L., Yoon, M., Motamedi, A., Kroll, A., et al. (2015). Reversible centriole depletion with an inhibitor of Polo-like kinase 4. *Science (New York, N. Y.), 348*, 1155–1160. https://doi.org/10.1126/SCIENCE.AAA5111.

CHAPTER THREE

Isolation, characterization, and experimental applications of apoptotic bodies *in vitro*

Atomu Yamaguchi[a,b,c,d], Alejandra García-Carrasco[a,c], Xingping Qin[a,c,d], and Kristopher Sarosiek[a,c,d,*]
[a]John B. Little Center for Radiation Sciences, Harvard T.H. Chan School of Public Health, Boston, MA, United States
[b]Department of Rehabilitation Science, Kobe University Graduate School of Health Sciences, Kobe, Hyogo, Japan
[c]Molecular and Integrative Physiological Sciences Program, Harvard T.H. Chan School of Public Health, Boston, MA, United States
[d]Laboratory of Systems Pharmacology, Harvard Program in Therapeutic Science, Department of Systems Biology, Harvard Medical School, Boston, MA, United States
*Corresponding author. e-mail address: sarosiek@hsph.harvard.edu

Contents

Methods in Cell Biology, Volume 210
ISSN 0091-679X, https://doi.org/10.1016/bs.mcb.2026.05.009

Abstract

Apoptotic bodies (ApoBDs) are a type of extracellular vesicle (EV) generated during the final stages of apoptosis. These vesicles were traditionally regarded as apoptotic cellular waste. More recently, they have begun to be recognized as mediators of intercellular communication via the delivery of diverse factors, with emerging roles in various physiological processes including immune regulation, cancer progression, and tissue repair. However, the lack of standardized methods for isolating and characterizing ApoBDs has limited both their study and potential application. Here, we present simple, efficient, and reproducible protocols for ApoBD isolation, characterization, and application to downstream experimental approaches. This protocol includes a two-step differential centrifugation to isolate ApoBDs and flow cytometry-based characterization, as well as optional steps such as caspase inhibitor-based modulation of ApoBD release and fluorescent labeling of ApoBDs for their tracking. Furthermore, we provide methods to evaluate ApoBD uptake by recipient cells, enabling subsequent downstream functional studies. These approaches promote the reproducibility and standardization of ApoBD research, providing a foundation to deepen our understanding of their functional roles and advance future clinical applications.

1. Introduction

Apoptosis (programmed cell death) is a biological process that plays critical roles not only in development and maintenance of tissue homeostasis, but also in a wide range of pathological conditions (Vitale et al., 2023). It can be triggered by various stimuli in the body, including responses of cancer or healthy cells to cancer therapeutics such as cytotoxic and targeted chemotherapeutic agents as well as radiation therapy (Sarosiek et al., 2013; Spetz et al., 2022), viral infections (Inde et al., 2021), and environmental exposures (Johnson & Sarosiek, 2024). A hallmark of apoptosis is the release of apoptotic bodies (ApoBDs), which are membrane-bound extracellular vesicles (EVs) that facilitate the clearance of dying cells in a non-inflammatory and regulated manner (Akers et al., 2013; Lynch et al., 2017). These vesicles may serve as an important mediator in tissue and organismal responses to apoptotic stimuli (Xu et al., 2019; Yu et al., 2023).

As apoptosis progresses, cells actively undergo a series of morphological changes, including cytoskeletal reorganization, nuclear fragmentation, and membrane blebbing, ultimately fragmenting into ApoBDs (Atkin-Smith & Poon, 2017; Yu et al., 2023). Specifically, activation of ROCK1 by caspase-3 induces actomyosin contraction and, together with PAK2 and

LIMK1, promotes apoptotic membrane blebbing (Coleman et al., 2001; Santavanond et al., 2021). This is followed by the opening of PANX1 channels by caspase-3 and 7, leading to the release of "find me" signals and cytoskeletal rearrangement (Chekeni et al., 2010; Poon et al., 2014). Eventually, apoptotic membrane protrusions are fragmented to form individual ApoBDs. These vesicles typically range from 1 to 5 μm in diameter and contain diverse intracellular components such as cytoplasmic proteins, lipids, fragmented DNA, RNA, and organelles including mitochondria (Karpman et al., 2017; Yu et al., 2023).

EVs are membrane-enclosed particles released by cells into the extracellular space and are commonly classified into exosomes, microvesicles, and ApoBDs according to their size and mode of biogenesis (Dang et al., 2020). Exosomes (50–150 nm) are generated within the endosome and released upon fusion of multivesicular bodies with the plasma membrane, whereas microvesicles (50–1000 nm) are formed by budding and fission of the plasma membrane. ApoBDs are large EVs that arise from membrane blebbing and fragmentation of apoptotic cells and typically range from 1000 to 5000 nm in diameter (Fig. 1). While these size ranges are widely used to distinguish EV subtypes, the size distributions of different EVs can partially overlap. In addition, novel subtypes of EVs have been reported, including blebbisomes, which are large organelle-rich extracellular vesicles (Jeppesen et al., 2025), and oncosomes, which are particularly large tumor-derived EVs (Minciacchi et al., 2015). In this protocol, we define ApoBDs

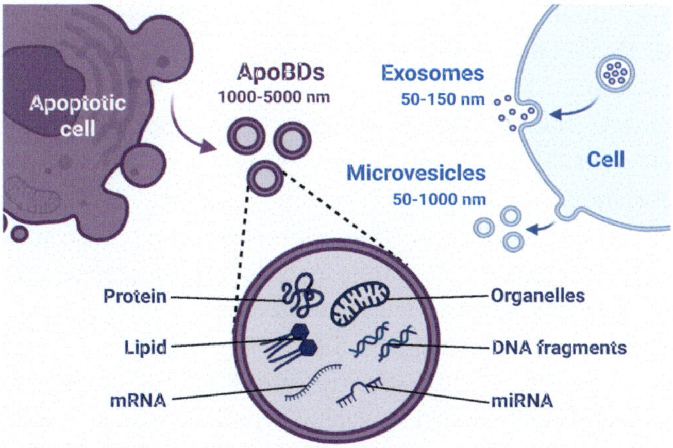

Fig. 1 Release of apoptotic bodies (ApoBDs) and other extracellular vesicles: microvesicles and exosomes from cells. *Created with Biorender.com.*

as relatively large, apoptosis-derived EVs that fall within the 1000–5000 nm range. A particularly notable feature of ApoBDs is the prominent externalization of phosphatidylserine (PS) on their membrane (Atkin-Smith & Poon, 2017). Although PS exposure can also occur on exosomes and microvesicles (Battistelli & Falcieri, 2020), it is consistently observed in ApoBDs, where it serves as an "eat me" signal to facilitate rapid recognition and clearance by phagocytic cells such as macrophages and dendritic cells (Ravichandran, 2011; Xiao et al., 2024). This process ensures their efficient, non-inflammatory removal, thereby contributing to the resolution of cell death and the maintenance of tissue homeostasis (Arandjelovic & Ravichandran, 2015). In addition, ApoBDs retain surface markers of their cells of origin (Atkin-Smith et al., 2017) and often encapsulate fragmented cellular contents such as nuclear DNA and mitochondria (Yu et al., 2023).

Beyond their traditional classification as cellular debris, ApoBDs have recently emerged as active mediators of intercellular communication with diverse biological functions and therapeutic potential (Caruso & Poon, 2018; Yu et al., 2023). Among their various potential biological roles, their involvement in cancer progression and immune modulation are their most well-documented functions. Depending on their cellular origin and cargo composition, ApoBDs can trigger distinct cellular responses, including the activation of cancer progression and suppression of immune responses (Li et al., 2022; Liu et al., 2020; Pavlyukov et al., 2018). The diagnostic and prognostic utility of ApoBDs is another promising area (Miao et al., 2024). Since ApoBDs preserve molecular signatures of their parent cells (Jiang et al., 2017), their content can serve as biomarkers for early disease detection or therapeutic response monitoring.

Prior research studies in ApoBDs have used a wide variety of apoptosis-inducing stimuli, isolation methods, and characterization strategies across different cell types. Table 1 summarizes representative reports in which the protocols for inducing apoptosis, isolating ApoBDs, and analyzing their properties are clearly described. Several protocols have been published for isolating and characterizing ApoBDs, including fluorescence-activated cell sorting (FACS)-based approaches and differential centrifugation strategies (Atkin-Smith et al., 2017; Jiang et al., 2017; Poon et al., 2019). FACS-based methods enable highly specific enrichment of cell type–defined ApoBDs but require access to specialized instruments and can be time-consuming, which may limit their implementation and scalability in many laboratories. In contrast, centrifugation-based protocols rely on widely available equipment and are relatively simple to perform, but they often

Table 1 Representative examples of apoptosis-inducing stimuli, Apoptotic body (ApoBD) isolation methods, and characterization strategies used in previous studies across different cell types.

Cell type	Apoptosis induction	ApoBD isolation method	ApoBD characterization method	References
THP-1, Jurkat T, HUVEC	UV	Centrifugation (3000 × *g*), FACS	Flow cytometry (AxV^+/TO-PRO3$^-$)	Atkin-Smith et al. (2017)
THP−1, Jurkat T, HL−60, HUVEC	UV, anti-Fas	Centrifugation (3000 × *g*)	Flow cytometry, Confocal microscopy (AxV^+/TO-PRO3$^-$)	Jiang et al. (2017)
THP−1, Jurkat T	UV	Centrifugation (3000 × *g*)	Flow cytometry (AxV^+/TO-PRO3$^-$)	Poon et al. (2019)
huMSC	UV	Centrifugation (12,000 × *g*)	TEM, WB (histone H3.3, H2B, C3b, and C1QC)	Xin et al. (2022)
MSC	STS	Centrifugation (16,000 × *g*)	Confocal microscopy (AxV^+), SEM, NTA	Liu et al. (2020)
Neutrophils	STS	Centrifugation (1000 × *g*)	Confocal microscopy (AxV^+/$C1q^+$), SEM, DLS	Bao et al. (2022)
Osteoclast	STS	Centrifugation (3000 × *g*)	Confocal microscopy (AxV^+), WB (C1QC, C3B, H3 and H2B)	Wu et al. (2023)
Osteoclast	STS	Centrifugation (3000 × *g*)	Flow cytometry (AxV^+/PI^-)	Ma et al. (2021)

(continued)

Table 1 Representative examples of apoptosis-inducing stimuli, Apoptotic body (ApoBD) isolation methods, and characterization strategies used in previous studies across different cell types. (*cont'd*)

Cell type	Apoptosis induction	ApoBD isolation method	ApoBD characterization method	References
MSC	Hypoxia	Centrifugation (2000 × *g*), filtration	Confocal microscopy ($AxV^+/TSP1^+$)	Li et al. (2023)
Endplate chondrocyte	H_2O_2	Centrifugation (3000 × *g*)	Confocal microscopy (AxV^+)	Yuan et al. (2019)
Osteoclast	Alendronate	Centrifugation (3500 × *g*), filtration	Flow cytometry (AxV^+/PI^-)	Ma et al. (2019)

Abbreviations: UV: Ultraviolet light, STS: Staurosporine, MSC: Mesenchymal stem cell, AxV: Annexin V, TEM: Transmission electron microscopy, WB: Western blot, PI: Propidium iodide, SEM: Scanning electron microscopy, NTA: Nanoparticle tracking analysis, DLS: Dynamic light scattering. THP-1, Jurkat T and HL-60 denote immortalized human cell lines, whereas all other cell types listed represent primary cell preparations.

differ in centrifugation parameters and lack standardized, quantitative criteria for ApoBD identification, making direct comparison across studies difficult. This methodological heterogeneity restricts the development of ApoBD-based therapeutics and diagnostics. Addressing this point is essential for unlocking the full potential of ApoBDs in both basic research and clinical applications.

In the present work, we build on these prior approaches by combining a standardized two-step low-speed centrifugation procedure with a robust annexin V/DAPI and scatter-based flow cytometry gating strategy that allows quantitative assessment of ApoBD yield and purity. We further incorporate optional modules, including caspase inhibition, fluorescent labeling, and uptake assays, to facilitate both mechanistic and functional studies. Together, these features are intended to promote methodological standardization and reproducibility while remaining broadly accessible. However, like other centrifugation-based approaches, our protocol cannot fully resolve ApoBDs from other large EVs with overlapping size and PS exposure and is currently optimized for *in vitro* cell culture rather than for complex biological fluids.

2. Materials

2.1 Common disposables (see *Note 1*)

- VWR® 12-well Cell Culture Plates (#10062-894, VWR)
- CELLSTAR® 96-well Tissue Culture Plates (#82050−634, Greiner Bio-One GmbH)
- WATSON™ 1.5 mL Microtube, Clear, with Snap Cap, Sterile (#131-8155C, WATSON Bio Lab)

2.2 Cells and reagents (see *Note 1*)

- Cell lines of interest (Any cell type can be used depending on the experimental purpose)
- Gibco™ RPMI 1640 Medium, with L-glutamine and phenol red, without HEPES (#11875119, Thermo Fisher Scientific)
- Gibco™ Fetal Bovine Serum, Value, heat-inactivated (#A5256801, Thermo Fisher Scientific)
- Gibco™ Penicillin-Streptomycin (10,000 U/mL), sterile-filtered (#15-140-163, Thermo Fisher Scientific)
- Staurosporine, ≥ 98 % (HPLC), powder (#HY-15141, MedChemExpress)

- Q-VD-OPh, ≥98 % (HPLC), powder (#HY-12305, MedChemExpress)
- Gibco™ Phosphate Buffered Saline (PBS), pH 7.4, 1X, sterile-filtered (#10010049, Thermo Fisher Scientific)
- Gibco™ Trypsin-EDTA (0.25 %), phenol red (#25200114, Thermo Fisher Scientific)
- Annexin V, Alexa Fluor™ 647 conjugate (#A23204, Thermo Fisher Scientific)
- DAPI Staining Solution (#ab228549, Abcam) (Alternative DNA dyes, including propidium iodide or TO-PRO-3 may be used as alternatives to DAPI, with a suitable annexin V conjugate chosen based on compatibility.)
- CFDA-SE (CFSE), ≥99 % (HPLC), powder (#HY-D0938, MedChemExpress)
- Hoechst 33342 Staining Dye Solution (#ab228551, Abcam)

2.3 Equipment (see *Note 1*)

- Humidified cell culture incubator, such as VWR Symphony CO_2 Water Jacketed Incubator (#3074, VWR)
- X-ray irradiator, such as RS 2000 X-ray Biological Irradiator (#RS-2000, Rad Source Technologies)
- Benchtop centrifuge, such as Eppendorf Centrifuge 5415 R, refrigerated microcentrifuge (#5415 R, Eppendorf)
- Flow cytometer equipped with lasers for acquisition of fluorescence from Alexa Fluor 647 and DAPI, such as Attune™ NxT Flow Cytometer (Thermo Fisher Scientific)

2.4 Software (see *Note 1*)

- Attune Flow Cytometer Software v 7.0 (Thermo Fisher Scientific)
- Prism v 10.2.3 (GraphPad)

3. Methods (*see Note 2*)

3.1 Cell culture and apoptosis induction

1. Seed 1–5 × 10^5 cells in 1.5 mL of cell culture medium in a 12 well plate and incubate overnight (*see Note 3*).
2. Discard the culture medium and add fresh medium, then induce apoptosis in the cells by treating with staurosporine (50–500 nM), by irradiation (2–10 Gy) or other treatment of choice (*see Note 4*) as well as vehicle or sham treatment for control wells.

3. Treat a subset of wells containing apoptosis-induced cells with the caspase inhibitor Q-VD-OPh (10–20 μM) to serve as a negative control.
4. Incubate the cells for 24–72 h, monitoring cell morphology and ApoBD release, and harvest at the time point that yields abundant ApoBDs. The time required for the release of ApoBDs varies depending on the cell type, the type of apoptotic stimulus, and its dose (*see Note 5*).

3.2 ApoBD isolation

1. Collect 1 mL of apoptotic conditioned medium in 1.5 mL microcentrifuge tubes. If using suspension cells, the cell suspension can simply be collected directly into a 1.5 mL microcentrifuge tube.
2. Centrifuge at 300 × *g* for 10 min at 4 °C.
3. Carefully collect 900 μL of the supernatant without disturbing the pellet and transfer it to a new tube.
4. Centrifuge at 3000 × *g* for 20 min at 4 °C (Fig. 2). To prevent drug carryover when using extracted ApoBDs for applications such as recipient cell treatment, the pellet may be washed with PBS after the 10 min centrifugation at 3000 × *g*, followed by a second centrifugation for 10 min
5. Remove all traces of fluid by aspiration and resuspend the pellet in 150 μL of PBS.
6. Isolated ApoBDs can be stored for 2–3 days at 4 °C for subsequent characterization.

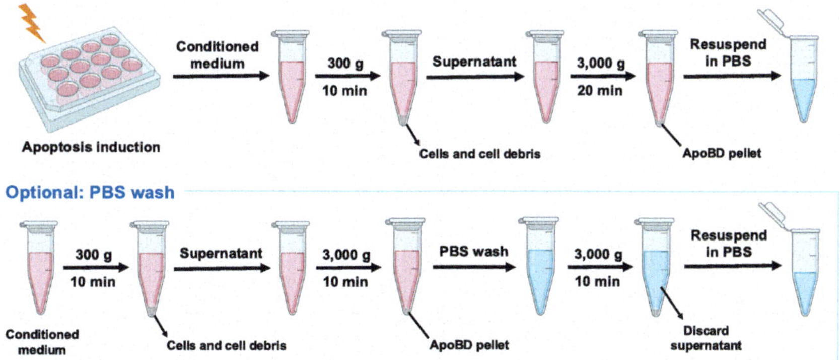

Fig. 2 Flowchart of apoptotic body (ApoBD) isolation protocol. Apoptotic cell culture supernatants were centrifuged at 300 × *g* for 10 min, followed by centrifugation of the supernatant at 3000 × *g* for 20 min. The resulting ApoBD pellet was resuspended in PBS. If required, the ApoBD pellet can be washed with PBS after the 10-min centrifugation at 3000 × *g*, and then centrifuged again for 10 min.

3.3 Staining and flow cytometry analysis

1. Prepare a 10 × staining solution by supplementing 10 × annexin V binding buffer with Alexa Fluor 647–Annexin V (1:100) and DAPI (1:500) (Annexin V binding buffer: 100 mM HEPES free acid, 40 mM KCl, 1.4 M NaCl, 7.5 mM, $MgCl_2$, 25 mM $CaCl_2$. Adjust pH to 7.4).
2. Transfer 100 μL of the resuspended ApoBD solution to a separate microcentrifuge tube. When using a flow cytometer with an autosampler, the sample can alternatively be transferred to a U-bottom 96-well plate.
3. Add 10 μL of the staining solution to each tube, protect from light, and incubate on ice for 30 min. Include an unstained control for setting gates during flow cytometry analysis.
4. Perform flow cytometry and confirm the induction of apoptosis by comparing non-apoptotic and apoptotic samples (Fig. 3).

3.4 Gating

1. Gate out DAPI-high events with reference to the non-centrifuged whole apoptotic sample (Fig. 4). This strategy removes any DNA-containing cell remnants that have a compromised plasma membrane.
2. Define Annexin V-positive events using the unstained control.
3. Gate FSC-low events based on the whole apoptotic sample.
4. Calculate the purity and concentration of the isolated ApoBDs using the following formula.

$$ApoBD\ purity = \frac{Number\ of\ gated\ ApoBDs}{Number\ of\ all\ events} \times 100$$

$$ApoBD\ concentration = \frac{Number\ of\ gated\ ApoBDs}{Volume\ used\ for\ ApoBD\ number\ measurement}$$

5. Confirm the increased production of ApoBDs in apoptosis-induced samples compared to the control, and the reduction in ApoBD yield in Q-VD-OPh–treated samples (Fig. 5).

3.5 Fluorescent labeling of ApoBDs

1. Seed $1-5 \times 10^5$ cells in 1.5 mL of cell culture medium in a 12 well plate and incubate overnight (*see Note 3*).
2. Wash the cells twice with PBS and add CFSE (5 μM final concentration in PBS), then incubate at 37 °C with 5 % CO_2 for 20 min

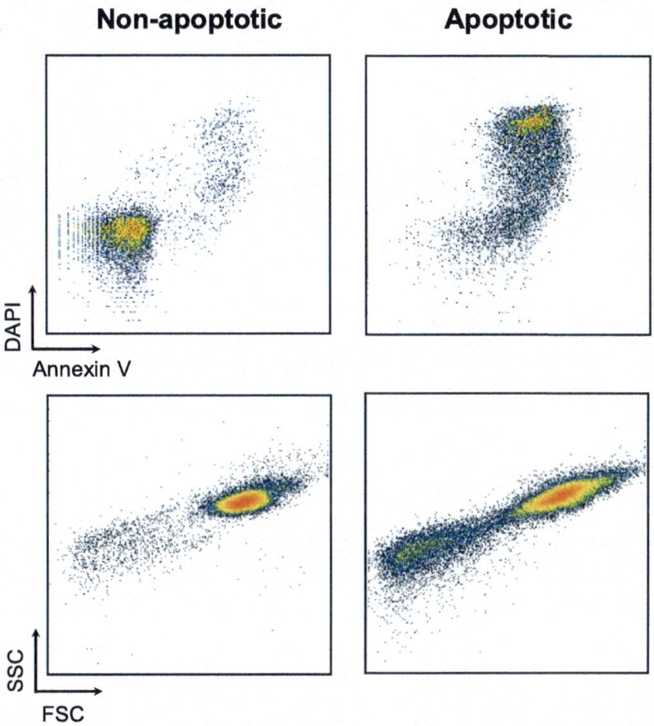

Fig. 3 Representative flow cytometry plots comparing non-apoptotic and apoptotic samples. THP-1 cells were treated with 250 nM staurosporine for 24 h. Apoptotic cells are characterized by increased Annexin V and DAPI positivity. In the forward scatter (FSC) versus side scatter (SSC) plot, the apoptotic sample showed a significant increase in a population smaller than the cell population, which largely represents apoptotic bodies.

3. Wash the cells twice with PBS, then incubate them in medium supplemented with FBS at 37 °C with 5 % CO_2 for at least 10 min. Avoid extended incubation, as it may result in CFSE signal dilution caused by cell proliferation. Move on to the next step within 3 h.
4. Stimulate the cells to induce apoptosis, as described in Section 3.1 and collect ApoBDs as described in Section 3.2.
5. Verify that the isolated ApoBDs are CFSE-positive using flow cytometry (Fig. 6) (*see Note 6*).

3.6 ApoBD uptake assay

1. Seed recipient cells into a 96-well plate and incubate for 24 h.
2. Apply CFSE-positive ApoBDs, isolated as described in Section 3.5, to recipient cells at a ratio of 1:1−10 (cells:ApoBDs), and co-incubate for 0−24 h.

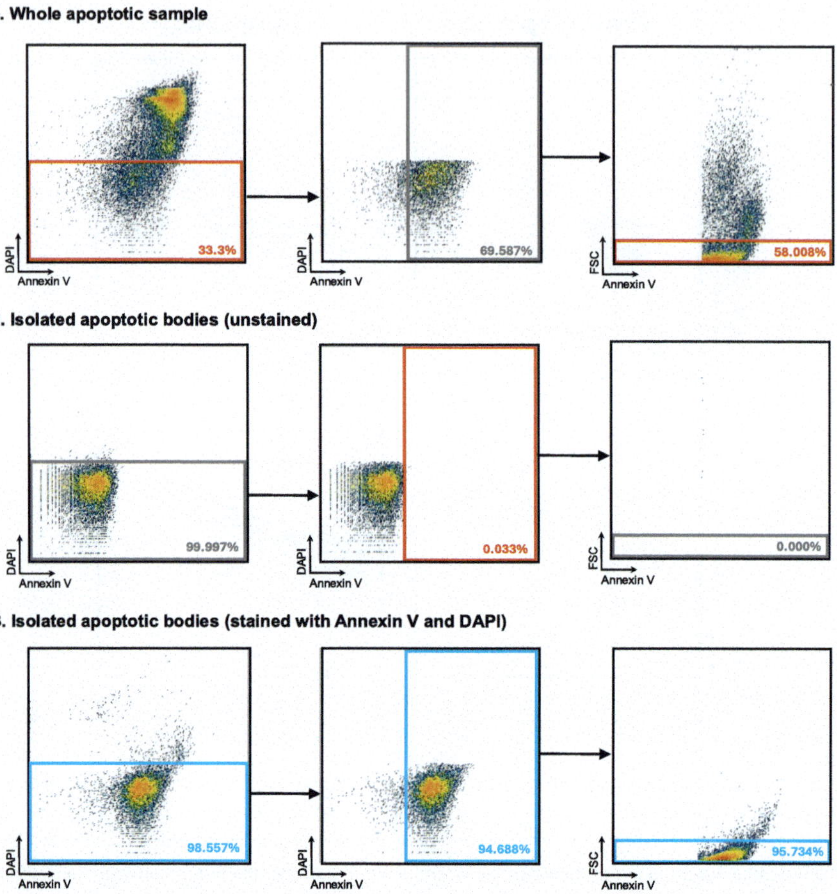

Fig. 4 Flowchart of apoptotic body (ApoBD) gating protocol. Apoptosis in THP-1 cells was induced by staurosporine, and ApoBDs were isolated by differential centrifugation 24 h later. The ApoBDs were gated as detailed in Section 3.4.

3. Remove the medium and wash the cells three times with PBS. For the 0 h time point, wash the cells immediately with PBS after the addition of ApoBDs without incubation. It is also recommended to prepare a separate control sample without ApoBD treatment.
4. Treat the cells with trypsin for 5 min to detach them and release cell-surface-bound ApoBDs, then add medium containing FBS to stop the reaction. Suspension cells should also be treated with trypsin to release surface-bound ApoBDs.
5. Perform flow cytometry analysis and gate the cell population based on FSC versus SSC plots (Fig. 7).

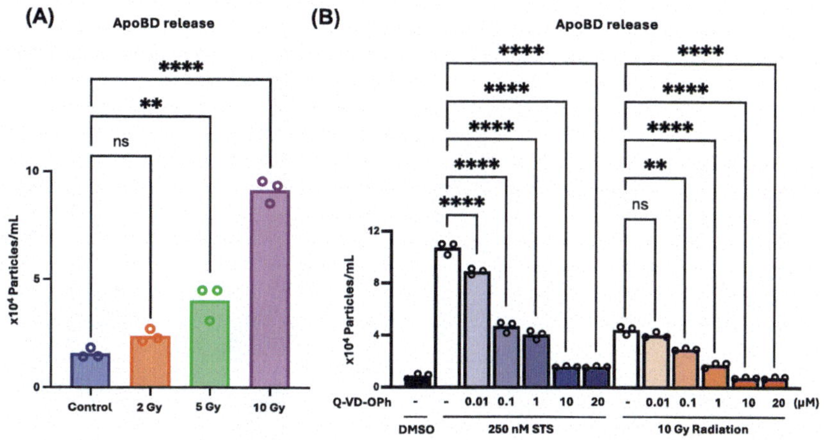

Fig. 5 (A) Dose-dependent release of apoptotic bodies (ApoBDs) induced by irradiation. THP-1 cells were exposed to 2 Gy, 5 Gy, or 10 Gy of radiation, and ApoBDs were isolated and quantified 48 h post-irradiation. (B) Caspase inhibition suppresses the formation of ApoBDs. THP-1 cells were treated with 250 nM STS or 10 Gy irradiation, followed by treatment with Q-VD-OPh. ApoBD release was quantified 24 h after treatment. Data are presented as mean ± SD of n = 3 biological replicates. One-way ANOVA followed by Tukey's multiple comparisons test was used; $p < 0.05$ was considered significant. **$p < 0.01$, ***$p < 0.001$, ****$p < 0.0001$.

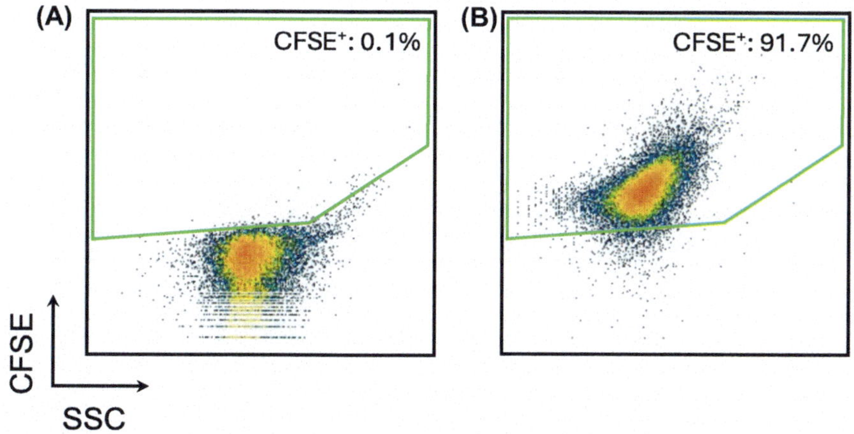

Fig. 6 Fluorescent Labeling of apoptotic bodies (ApoBDs) with CFSE. ApoBDs were isolated from OVCAR4 cells that were treated with STS and either no stain (A) or stained with CFSE (B). Percentages of $CFSE^+$ ApoBDs are calculated. Over 90 % of ApoBDs released from CFSE-labeled OVCAR4 cells retained detectable CFSE signal.

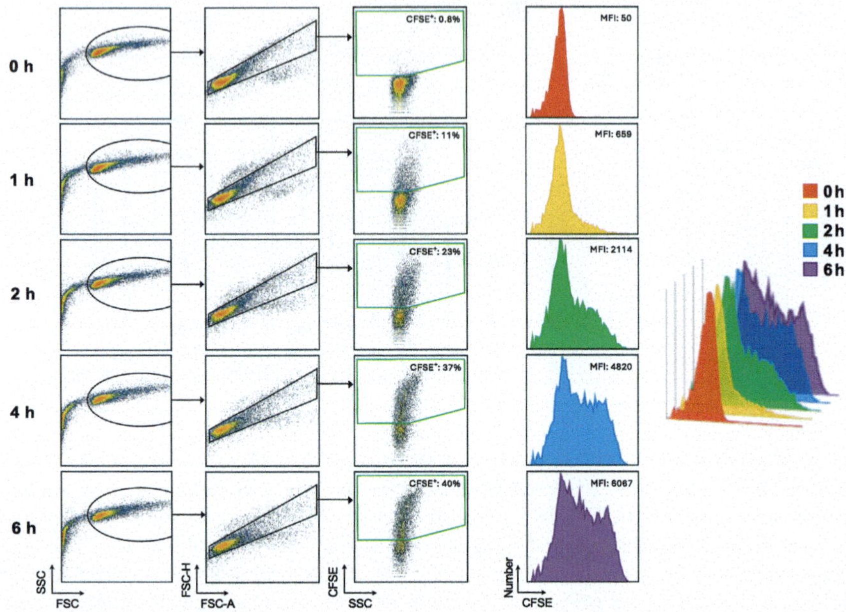

Fig. 7 Uptake of apoptotic bodies (ApoBDs) by macrophages. THP-1 monocytes were differentiated into macrophages by treatment with 80 ng/mL phorbol 12-myristate 13-acetate for 24 h followed by 24 h incubation in complete medium, then co-cultured with CFSE-positive ApoBDs derived from OVCAR4 cells. The uptake of ApoBDs was assessed after 0–6 h of co-incubation.

6. Select singlets using FSC-A versus FSC-H plots.
7. Identify cells that have engulfed ApoBDs by gating for CFSE$^+$ events.
8. Quantify the proportion of cells that have taken up ApoBDs at each time point.

3.7 Troubleshooting

1. **Low ApoBD yield:** Poor ApoBD production is often due to insufficient induction of apoptosis. We recommend first confirming efficient apoptosis induction by assessing annexin V/DAPI staining in the cells and, if necessary, increasing the concentration or treatment duration of the apoptotic stimulus in pilot experiments. As ApoBD yield strongly depends on the type and dose of stimulus, switching to another stimulus (e.g., from irradiation to STS) or increasing the stimulus dose can enhance ApoBD production.
2. **Poor annexin V staining of ApoBDs:** Efficient PS detection requires the presence of Ca^{2+} in the binding buffer and appropriate temperature

and incubation times. Make sure to use the recommended annexin V binding buffer and avoid EDTA-containing solutions.

3. **Weak fluorescent labeling of ApoBDs:** As CFSE fluorescence is progressively diluted as cells divide, it is important to avoid long delays between cell labeling and induction of apoptosis. We recommend inducing apoptosis within approximately 3 h after CFSE staining to maintain a sufficiently strong signal in ApoBDs. If high background fluorescence is observed in the supernatant, cells should be washed several times with PBS or culture medium after CFSE staining and before apoptosis induction to remove excess unbound dye.

4. Notes

1. Virtually equivalent disposables, reagents, equipment and software are available from a variety of alternative providers.
2. All experiments in this protocol were performed in at least $n = 3$ independent biological replicates, and key steps of the protocol (apoptosis induction, ApoBD isolation and flow cytometric analysis) were validated in multiple cell lines (THP-1, HeLa, OVCAR3, OVCAR4, U-2 OS, MDA-MB-231, and OCI-AML2) to confirm reproducibility across cell types.
3. Seeding $1–5 \times 10^5$ cells typically results in ApoBD yields on the order of $10^4–10^5$ collected ApoBDs, depending on the cell type, stimulus, dose, and treatment duration. To increase the yield of ApoBDs, a higher number of cells can be seeded in larger culture dishes or flasks. Under the standard conditions described in this protocol yields sufficient ApoBDs for flow cytometry–based characterization, uptake assays and extraction of RNA from ApoBDs for targeted gene expression analyses. In contrast, more material-intensive applications such as proteomics require a substantial up-scaling of the starting material (e.g. multiple large culture flasks or equivalent).
4. Note that ApoBDs may potentially retain agents that were used to induce apoptosis and may transmit those agents to recipient cells.
5. It is recommended to monitor cells by microscopy and/or flow cytometry at several time points between 24 and 72 h. As a practical guideline, ApoBD harvest is optimal when most cells show clear apoptotic morphology (cell rounding, membrane blebbing, and/or detachment) and abundant small particles are visible in the supernatant.

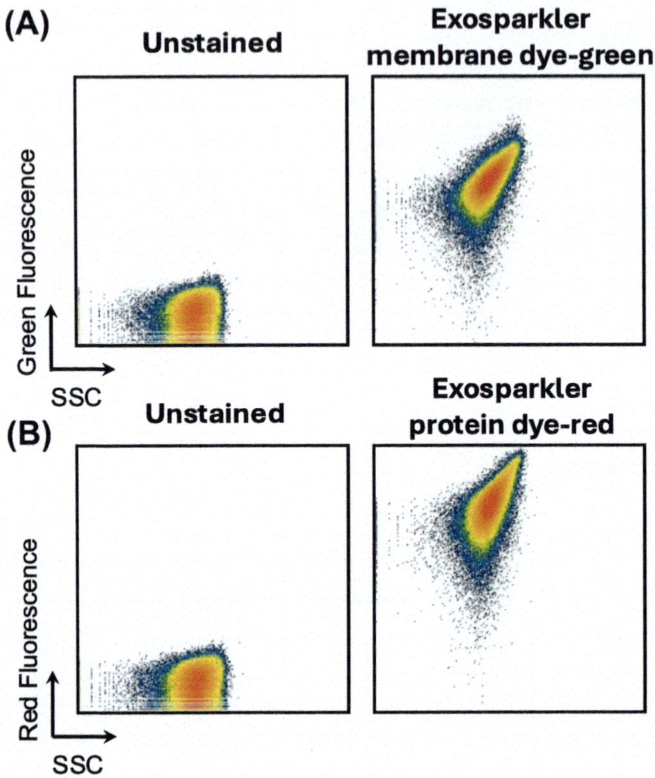

Fig. 8 Fluorescent labeling of apoptotic bodies (ApoBDs) with (A) the exosparkler exosome membrane labeling kit (Green) or (B) the exosparkler exosome protein labeling kit (Red). ApoBDs were isolated from THP-1 cells that were treated with STS and stained following the manufacturer's instructions.

6. Alternatively, the Exosparkler Exosome Membrane Labeling Kit (EX01-10, Dojindo Laboratories) or Exosparkler Exosome Protein Labeling Kit (EX05-10, Dojindo Laboratories) can be used to fluorescently label ApoBDs after isolation (Fig. 8).

5. Concluding remarks

Here, we provide a simple and reproducible protocol that encompasses ApoBD isolation, analysis, fluorescent labeling, and the uptake assay by recipient cells. Previous studies employed a wide range of methods for the isolation and characterization of ApoBDs, making it difficult to compare results across different studies and ensure reproducibility. While the two-step, low-speed centrifugation protocol is used in some previous studies (Atkin-Smith et al., 2017; Jiang et al., 2017; Poon et al., 2019), we have optimized it by incorporating an optional washing step to minimize drug carryover depending on downstream applications. We also established a robust gating strategy, enabling more accurate identification of ApoBDs. Furthermore, optional modules such as fluorescent labeling and caspase inhibition enhance the versatility of our approach. These advances enable comprehensive monitoring of ApoBD dynamics, from their release to uptake by recipient cells, thereby facilitating their functional evaluation.

Although ApoBDs were historically regarded as waste after cell death without any functions, recent studies have highlighted their biological roles and relevance to various disease contexts (Xu et al., 2019). These vesicles retain a diverse array of parent cell components and are increasingly recognized as functional carriers that can influence neighboring cells (Yu et al., 2023). Recent findings indicate their roles in cancer progression (Pavlyukov et al., 2018), immune regulation (Caruso & Poon, 2018), and tissue repair (Liu et al., 2020). Nevertheless, the current understanding of the functional roles and applications of ApoBDs remains very limited.

This ApoBD isolation protocol does not require specialty instruments or antibodies, making it broadly accessible to a wide range of researchers. Furthermore, this approach can be employed with any type of cells or apoptotic stimuli, allowing for systematic comparison of ApoBDs under multiple biological and pathological conditions. While the protocol described here was optimized and validated for the isolation of ApoBDs (1000–5000 nm) released into cell culture supernatants, we have not yet evaluated its performance for tissue- or biofluid-derived samples, for which some published protocols employ higher centrifuge forces or ultracentrifugation-based steps, or for the isolation of smaller apoptotic EVs. Altogether, this approach offers a robust and reproducible platform for in vitro ApoBD studies and facilitates the exploration of their broader experimental and clinical potential. Future studies utilizing these protocols are expected to deepen our understanding of the physiological and pathological significance of ApoBDs.

Acknowledgments

We thank the members of our labs for their thoughtful suggestions on this work.

Funding

This work was supported by funding from:
JSPS Overseas Research Fellowships (A.Y.).
JSPS Grant-in-Aid for Early-Career Scientists 25K20839 (A.Y.).
JSPS Fund for the Promotion of Joint International Research 23KK0305 (A.Y.).
Takeda Science Foundation (A.Y.).
Alex's Lemonade Stand Foundation for Childhood Cancers (K.A.S.).
Andrew McDonough B + Foundation (K.A.S.).
HSPH Dean's Fund for Scientific Advancement (K.A.S.).
National Institutes of Health grant R37CA248565 (K.A.S.).
National Institutes of Health grant R01DK125263 (K.A.S.).
Ovarian Cancer Research Alliance grant CRDG-2023-3-1001 (K.A.S.).
Zhu Family Fund (K.A.S.).

Competing interests

All authors declare no competing interests.

References

Akers, J. C., Gonda, D., Kim, R., Carter, B. S., & Chen, C. C. (2013). Biogenesis of extracellular vesicles (EV): Exosomes, microvesicles, retrovirus-like vesicles, and apoptotic bodies. *Journal of Neuro-oncology, 113*(1), 1–11. https://doi.org/10.1007/s11060-013-1084-8.

Arandjelovic, S., & Ravichandran, K. S. (2015). Phagocytosis of apoptotic cells in homeostasis. *Nature Immunology, 16*(9), 907–917. https://doi.org/10.1038/ni.3253.

Atkin-Smith, G. K., Paone, S., Zanker, D. J., Duan, M., Phan, T. K., Chen, W., ... Poon, I. K. H. (2017). Isolation of cell type-specific apoptotic bodies by fluorescence-activated cell sorting. *Scientific Reports,* 7(1), 39846. https://doi.org/10.1038/srep39846.

Atkin-Smith, G. K., & Poon, I. K. H. (2017). Disassembly of the dying: Mechanisms and functions. *Trends in Cell Biology, 27*(2), 151–162. https://doi.org/10.1016/j.tcb.2016.08.011.

Bao, L., Dou, G., Tian, R., Lv, Y., Ding, F., Liu, S., ... Jin, Y. (2022). Engineered neutrophil apoptotic bodies ameliorate myocardial infarction by promoting macrophage efferocytosis and inflammation resolution. *Bioactive Materials, 9*, 183–197. https://doi.org/10.1016/j.bioactmat.2021.08.008.

Battistelli, M., & Falcieri, E. (2020). Apoptotic bodies: Particular extracellular vesicles involved in intercellular communication. *Biology, 9*(1), 21. https://doi.org/10.3390/biology9010021.

Caruso, S., & Poon, I. K. H. (2018). Apoptotic cell-derived extracellular vesicles: More than just debris. *Frontiers in Immunology, 9*, 1486. https://doi.org/10.3389/fimmu.2018.01486.

Chekeni, F. B., Elliott, M. R., Sandilos, J. K., Walk, S. F., Kinchen, J. M., Lazarowski, E. R., ... Ravichandran, K. S. (2010). Pannexin 1 channels mediate "find-me" signal release and membrane permeability during apoptosis. *Nature, 467*(7317), 863–867. https://doi.org/10.1038/nature09413.

Coleman, M. L., Sahai, E. A., Yeo, M., Bosch, M., Dewar, A., & Olson, M. F. (2001). Membrane blebbing during apoptosis results from caspase-mediated activation of ROCK I. *Nature Cell Biology, 3*(4), 339–345. https://doi.org/10.1038/35070009.

Dang, X. T. T., Kavishka, J. M., Zhang, D. X., Pirisinu, M., & Le, M. T. N. (2020). Extracellular vesicles as an efficient and versatile system for drug delivery. *Cells, 9*(10), 2191. https://doi.org/10.3390/cells9102191.

Inde, Z., Croker, B. A., Yapp, C., Joshi, G. N., Spetz, J., Fraser, C., ... Sarosiek, K. A. (2021). Age-dependent regulation of SARS-CoV-2 cell entry genes and cell death programs correlates with COVID-19 severity. *Science Advances, 7*(34), eabf8609. https://doi.org/10.1126/sciadv.abf8609.

Jeppesen, D. K., Sanchez, Z. C., Kelley, N. M., Hayes, J. B., Ambroise, J., Koory, E. N., ... Burnette, D. T. (2025). Blebbisomes are large, organelle-rich extracellular vesicles with cell-like properties. *Nature Cell Biology, 27*(3), 438–448. https://doi.org/10.1038/s41556-025-01621-0.

Jiang, L., Paone, S., Caruso, S., Atkin-Smith, G. K., Phan, T. K., Hulett, M. D., & Poon, I. K. H. (2017). Determining the contents and cell origins of apoptotic bodies by flow cytometry. *Scientific Reports, 7*(1), 14444. https://doi.org/10.1038/s41598-017-14305-z.

Johnson, L., & Sarosiek, K. A. (2024). Role of intrinsic apoptosis in environmental exposure health outcomes. *Trends in Molecular Medicine, 30*(1), 56–73. https://doi.org/10.1016/j.molmed.2023.11.003.

Karpman, D., Ståhl, A., & Arvidsson, I. (2017). Extracellular vesicles in renal disease. *Nature Reviews Nephrology, 13*(9), 545–562. https://doi.org/10.1038/nrneph.2017.98.

Li, J., Wei, C., Yang, Y., Gao, Z., Guo, Z., & Qi, F. (2022). Apoptotic bodies extracted from adipose mesenchymal stem cells carry microRNA-21–5p to induce M2 polarization of macrophages and augment skin wound healing by targeting KLF6. *Burns: Journal of the International Society for Burn Injuries, 48*(8), 1893–1908. https://doi.org/10.1016/j.burns.2021.12.010.

Li, X., Jiang, Y., Liu, X., Fu, J., Du, J., Luo, Z., ... Guo, L. (2023). Mesenchymal stem cell-derived apoptotic bodies alleviate alveolar bone destruction by regulating osteoclast differentiation and function. *International Journal of Oral Science, 15*(1), 51. https://doi.org/10.1038/s41368-023-00255-y.

Liu, H., Liu, S., Qiu, X., Yang, X., Bao, L., Pu, F., ... Jin, Y. (2020). Donor MSCs release apoptotic bodies to improve myocardial infarction via autophagy regulation in recipient cells. *Autophagy, 16*(12), 2140–2155. https://doi.org/10.1080/15548627.2020.1717128.

Liu, J., Qiu, X., Lv, Y., Zheng, C., Dong, Y., Dou, G., ... Jin, Y. (2020). Apoptotic bodies derived from mesenchymal stem cells promote cutaneous wound healing via regulating the functions of macrophages. *Stem Cell Research & Therapy, 11*(1), 507. https://doi.org/10.1186/s13287-020-02014-w.

Lynch, C., Panagopoulou, M., & Gregory, C. D. (2017). Extracellular vesicles arising from apoptotic cells in tumors: Roles in cancer pathogenesis and potential clinical applications. *Frontiers in Immunology, 8*, 1174. https://doi.org/10.3389/fimmu.2017.01174.

Ma, Q., Liang, M., Wu, Y., Ding, N., Duan, L., Yu, T., ... Dou, C. (2019). Mature osteoclast–derived apoptotic bodies promote osteogenic differentiation via RANKL-mediated reverse signaling. *Journal of Biological Chemistry, 294*(29), 11240–11247. https://doi.org/10.1074/jbc.RA119.007625.

Ma, Q., Liang, M., Wu, Y., Luo, F., Ma, Z., Dong, S., ... Dou, C. (2021). Osteoclast-derived apoptotic bodies couple bone resorption and formation in bone remodeling. *Bone Research, 9*(1), 5. https://doi.org/10.1038/s41413-020-00121-1.

Miao, X., Wu, X., You, W., He, K., Chen, C., Pathak, J. L., & Zhang, Q. (2024). Tailoring of apoptotic bodies for diagnostic and therapeutic applications:advances, challenges, and prospects. *Journal of Translational Medicine, 22*(1), 810. https://doi.org/10.1186/s12967-024-05451-w.

Minciacchi, V. R., You, S., Spinelli, C., Morley, S., Zandian, M., Aspuria, P.-J., ... Di Vizio, D. (2015). Large oncosomes contain distinct protein cargo and represent a separate functional class of tumor-derived extracellular vesicles. *Oncotarget, 6*(13), 11327–11341. https://doi.org/10.18632/oncotarget.3598.

Pavlyukov, M. S., Yu, H., Bastola, S., Minata, M., Shender, V. O., Lee, Y., ... Nakano, I. (2018). Apoptotic cell-derived extracellular vesicles promote malignancy of glioblastoma via intercellular transfer of splicing factors. *Cancer Cell, 34*(1), 119–135.e10. https://doi.org/10.1016/j.ccell.2018.05.012.

Poon, I. K. H., Chiu, Y.-H., Armstrong, A. J., Kinchen, J. M., Juncadella, I. J., Bayliss, D. A., & Ravichandran, K. S. (2014). Unexpected link between an antibiotic, pannexin channels and apoptosis. *Nature, 507*(7492), 329–334. https://doi.org/10.1038/nature13147.

Poon, I. K. H., Parkes, M. A. F., Jiang, L., Atkin-Smith, G. K., Tixeira, R., Gregory, C. D., ... Baxter, A. A. (2019). Moving beyond size and phosphatidylserine exposure: Evidence for a diversity of apoptotic cell-derived extracellular vesicles in vitro. *Journal of Extracellular Vesicles, 8*(1), 1608786. https://doi.org/10.1080/20013078.2019.1608786.

Ravichandran, K. S. (2011). Beginnings of a good apoptotic meal: The find-me and eat-me signaling pathways. *Immunity, 35*(4), 445–455. https://doi.org/10.1016/j.immuni.2011.09.004.

Santavanond, J. P., Rutter, S. F., Atkin-Smith, G. K., & Poon, I. K. H. (2021). Apoptotic bodies: Mechanism of formation, isolation and functional relevance. In S. Mathivanan, P. Fonseka, C. Nedeva, & I. Atukorala (Vol. Eds.), *New frontiers: Extracellular vesicles. vol. 97. New frontiers: Extracellular vesicles* (pp. 61–88). Springer International Publishing. https://doi.org/10.1007/978-3-030-67171-6_4.

Sarosiek, K. A., Ni Chonghaile, T., & Letai, A. (2013). Mitochondria: Gatekeepers of response to chemotherapy. *Trends in Cell Biology, 23*(12), 612–619. https://doi.org/10.1016/j.tcb.2013.08.003.

Spetz, J. K. E., Florido, M. H. C., Fraser, C. S., Qin, X., Choiniere, J., Yu, S. J., ... Sarosiek, K. A. (2022). Heightened apoptotic priming of vascular cells across tissues and life span predisposes them to cancer therapy-induced toxicities. *Science Advances, 8*(45), eabn6579. https://doi.org/10.1126/sciadv.abn6579.

Vitale, I., Pietrocola, F., Guilbaud, E., Aaronson, S. A., Abrams, J. M., Adam, D., ... Galluzzi, L. (2023). Apoptotic cell death in disease—Current understanding of the NCCD 2023. *Cell Death & Differentiation, 30*(5), 1097–1154. https://doi.org/10.1038/s41418-023-01153-w.

Wu, Y., Ai, H., Xi, Y., Yin, P., Qu, Y., Xu, J., ... Luo, F. (2023). Reduced osteoclast-derived apoptotic bodies in bone marrow characterizes the pathological progression of osteoporosis. *Cell Death Discovery, 9*(1), 135. https://doi.org/10.1038/s41420-023-01434-w.

Xiao, L., Zhang, L., Guo, C., Xin, Q., Gu, X., Jiang, C., & Wu, J. (2024). "Find Me" and "Eat Me" signals: Tools to drive phagocytic processes for modulating antitumor immunity. *Cancer Communications (London, England), 44*(7), 791–832. https://doi.org/10.1002/cac2.12579.

Xin, L., Wei, C., Tong, X., Dai, Y., Huang, D., Chen, J., ... Zhang, S. (2022). In situ delivery of apoptotic bodies derived from mesenchymal stem cells via a hyaluronic acid hydrogel: A therapy for intrauterine adhesions. *Bioactive Materials, 12*, 107–119. https://doi.org/10.1016/j.bioactmat.2021.10.025.

Xu, X., Lai, Y., & Hua, Z.-C. (2019). Apoptosis and apoptotic body: Disease message and therapeutic target potentials. *Bioscience Reports, 39*(1), BSR20180992. https://doi.org/10.1042/BSR20180992.

Yu, L., Zhu, G., Zhang, Z., Yu, Y., Zeng, L., Xu, Z., ... Pathak, J. L. (2023). Apoptotic bodies: Bioactive treasure left behind by the dying cells with robust diagnostic and therapeutic application potentials. *Journal of Nanobiotechnology, 21*(1), 218. https://doi.org/10.1186/s12951-023-01969-1.

Yuan, F., Xu, R., Ye, J., Zhao, M., Ren, L., & Li, X. (2019). Apoptotic bodies from endplate chondrocytes enhance the oxidative stress-induced mineralization by regulating PPi metabolism. *Journal of Cellular and Molecular Medicine, 23*(5), 3665–3675. https://doi.org/10.1111/jcmm.14268.

CHAPTER FOUR

Lipid ROS measurement by spectral flow cytometry based on the ratiometric dye BODIPY™ C11 581/591

Daniel Butcher[a,*], Naiqi Wang[a], Sam Nettelfield[a], and Di Yu[a,b,*]
[a]Faculty of Health, Frazer Institute, Medicine and Behavioural Sciences, University of Queensland, Brisbane, QLD, Australia
[b]Faculty of Health, Ian Frazer Centre for Children's Immunotherapy Research, Children's Health Research Centre, Medicine and Behavioural Sciences, University of Queensland, Brisbane, QLD, Australia
*Corresponding authors. e-mail address: d.butcher@uq.edu.au; di.yu@uq.edu.au

Contents

Abstract

The detection and measurement of lipid reactive oxygen species (ROS) is a critical method widely applied in the study of ferroptosis, a recently defined form of iron-dependent programmed cell death characterized by excessive lipid peroxidation and subsequent membrane damage. As ferroptosis research expands into various pathophysiological processes and human diseases, including cancer, immunity and inflammation, and tissue injuries, there is an increasing need for robust methods to measure lipid ROS in heterogenous cell samples, such as in deep immune cell profiling. BODIPY™ C11 581/591 is a gold-standard ratiometric fluorescent dye that reports the levels of lipid ROS, and has been used in both mechanistic and applied studies demonstrating ferroptosis regulation across multiple cell types. However, its

Methods in Cell Biology, Volume 210
ISSN 0091-679X, https://doi.org/10.1016/bs.mcb.2026.05.017

use has largely been confined to traditional flow cytometry due to its extensive spillover into multiple emission channels and its ratiometric properties, which complicate its application in spectral flow cytometry. Herein, we present solutions to these challenges and a comprehensive workflow for the optimization, unmixing, staining, and analysis required to multiplex lipid ROS detection using BODIPY™ C11 581/591 with high-parameter spectral flow cytometry panels. Our unmixing strategy represents a novel method for separating the fluorescent spectra of two forms of a ratiometric dye, which may also be applicable to other ratiometric dyes in spectral flow cytometry.

1. Introduction

Lipid reactive oxygen species (ROS) are essential biomolecules with critical roles in a diverse array of physiological and pathophysiological processes, ranging from cell signalling to cell death. With the insurgence of research into the cell death pathway of ferroptosis in the past decade, exploration into the lethal and non-lethal mechanisms by which lipid ROS can regulate biological systems has become a major research focus (Dixon & Pratt, 2023; Jin et al., 2024; Liu et al., 2025). According to the current paradigm, the accumulation of lipid ROS represents the universal core mechanism of ferroptosis, and in the absence of other consistent and robust markers of cell death by ferroptosis, detection of lipid ROS remains a critical method to identify ferroptosis in cells (Mishima et al., 2025; Stockwell, 2022; Wiernicki et al., 2020).

While much of the early work on lipid ROS and ferroptosis has been performed in the context of cancer cells (Doll et al., 2017; Yang et al., 2014; Yang & Stockwell, 2008), a substantial body of work has since identified crucial roles for ferroptosis in regulating non-cancer cells, particularly various subsets of immune cells (Liu et al., 2025; Ma et al., 2021; Matsushita et al., 2015; Morgan et al., 2024; Xu et al., 2021; Xu et al., 2021; Yao et al., 2021). To facilitate efficient exploratory research into ferroptosis in a wide variety of cell types, especially in the context of immunology, methods of detecting lipid ROS must be made compatible with techniques to assay multiple cell types simultaneously, such as flow cytometry.

The current gold-standard methods for detecting lipid ROS levels in cells include oxidative lipidomics or fluorescent dyes such as BODIPY™ C11 581/591 (henceforth, BODIPY C11) (Mishima et al., 2025; Stockwell, 2022). While bulk oxidative lipidomics has advantages in terms of sensitivity and level of detail in its results, it is currently limited by the need for a homogenous cell population to be assayed, precluding it from the simultaneous detection of lipid ROS across multiple cell subsets. By comparison, BODIPY C11 can be multiplexed with other fluorescent dyes and antibodies in flow cytometry panels to allow lipid ROS detection in multiple cell subsets within heterogenous samples. Therefore, BODIPY C11 is very well-suited to exploratory research, and has been widely used in traditional flow cytometry for measuring lipid ROS (Ma et al., 2021; Xu et al., 2021; Yao et al., 2021). However, the dye exhibits a broad emission spectrum, leading to extensive spillover into many other emission channels in traditional flow cytometry. This has precluded the use of BODIPY C11 in large high-parameter panels, such as those used in deep immune cell subset profiling.

The advent of spectral flow cytometry technology promises significant advancements on traditional flow cytometry in terms of panel size and staining resolution (Astakhova et al., 2025), which notably has the potential to facilitate high-parameter flow cytometry panels multiplexed with BODIPY C11 to detect lipid ROS. However, significant challenges exist due to BODIPY C11's ratiometric properties, whereby the dye changes fluorescence from a signature similar to phycoerythrin (PE) in its reduced form to a fluorescent signature similar to fluorescein isothiocyanate (FITC) in its oxidised form (henceforth referred to as BODIPY C11 RED and BODIPY C11 OX, respectively, short for 'Reduced' and 'Oxidised' forms; Fig. 1A). BODIPY C11 undergoes this fluorescent shift from BODIPY C11 RED to BODIPY C11 OX in the presence of lipid ROS, meaning that measurement of lipid ROS depends upon accurately unmixing both fluorescent signatures, and can be represented by the expression '$\frac{\textit{Intensity of BODIPY C11 OX}}{\textit{Intensity of BODIPY C11 RED}}$' (Dojindo, 2023).

In an effort to generate the single colour controls (SCCs) necessary to perform unmixing for spectral flow cytometry, we have tested a wide range of strategies to generate lipid ROS and produce cells exhibiting only the BODIPY C11 OX signature. However, regardless of how much oxidative stress is applied, cells stained with BODIPY C11 always retain a very bright signature of BODIPY C11 RED in addition to BODIPY C11 OX (Fig. 1B), and it is therefore impossible to generate true SCCs for unmixing in spectral flow cytometry. At present, there are no existing protocols or guidelines for unmixing ratiometric dyes like BODIPY C11 in spectral flow cytometry, particularly when a true SCC cannot be generated. Moreover, while the compensation algorithm for traditional flow cytometry is sufficiently simple

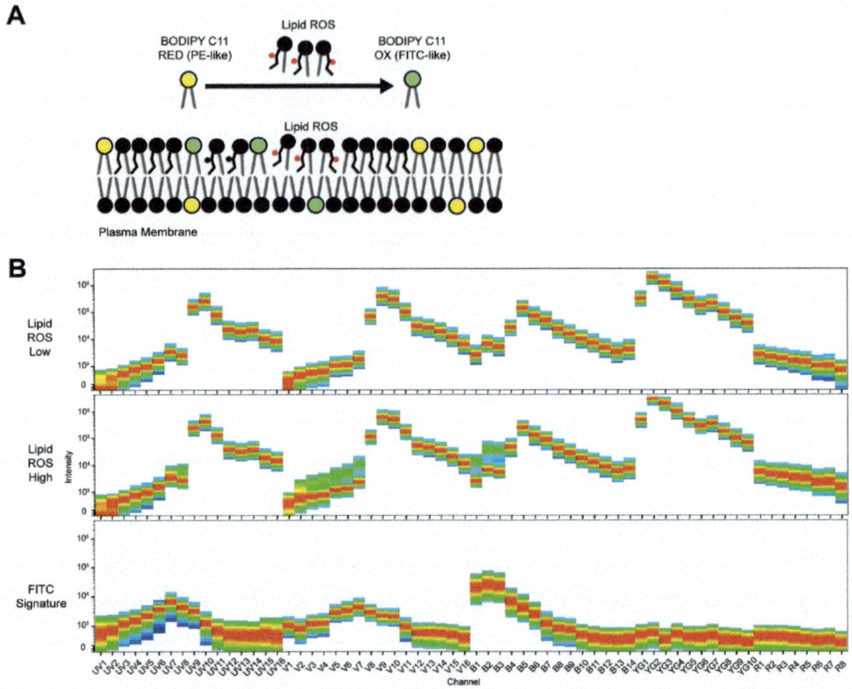

Fig. 1 Staining principle and flow cytometry spectra of BODIPY C11. A – Diagram representing the mechanism by which BODIPY C11 measures lipid ROS in cells. The dye begins in reduced form (emitting the BODIPY C11 RED signature), integrates into the cell membrane and becomes oxidised by local lipid ROS, changing fluorescence to its oxidised form (emitting the BODIPY C11 OX signature). B – Representative raw spectral flow cytometry 'signatures' of BODIPY C11 RED (lipid ROS low sample), BODIPY C11 OX (lipid ROS high sample), and a generic FITC-conjugated antibody for comparison. The 'signatures' show the intensity of the dye (y-axis) in each detector (x-axis) across the full spectrum measured by the flow cytometer. The lipid ROS high sample contains both the BODIPY C11 RED and BODIPY C11 OX signature within it, as the dye cannot be completely converted to the BODIPY C11 OX signature. The BODIPY C11 RED signature peaks in YG2, while the BODIPY C11 OX and FITC signatures both peak in B2.

that a surrogate SCC using a FITC-labelled antibody can be used in place of a true BODIPY C11 OX SCC, the complexity of the unmixing algorithm in spectral flow cytometry requires a more elegant solution.

Herein, we present a novel solution to these challenges to create a simply reproducible and customisable method to measure lipid ROS in high-parameter spectral flow cytometry. Rather than generating a true SCC, we instead generate 'lipid ROS high' and 'lipid ROS low' reference controls, and carefully select the 'negative' for each fluorescent tag to allow unmixing of BODIPY C11 OX from the BODIPY C11 RED signature. Crucially, these challenges with unmixing BODIPY C11 as a ratiometric dye are likely to arise when attempting to use other ratiometric dyes in spectral flow cytometry, and we anticipate that this present methodology will be applicable to unmix other ratiometric dyes as well.

The present protocol was generated using mouse splenocytes (with a focus on T cells), and the methods provided herein refer to staining splenocytes for flow cytometry. However, the information and methods are generalizable to any single cell suspension that can be analysed by flow cytometry.

2. Materials

2.1 Common disposables

- 96-well Clear Round Bottom Tissue Culture-treated Microplate (Corning®, #3799)
- Countess™ Cell Counting Chamber Slides with Trypan Blue (Invitrogen™, C10283)

2.2 Cells and reagents

- 1S,3R-RSL3 (RSL3, Sigma-Aldrich®, SML2234)
- 2-Mercaptoethanol (Gibco™, #21985023)
- Antibodies for flow cytometry, such as BUV496 rat anti-mouse CD4 (GK1.5, BD Biosciences, #612952), or Spark NIR™ 685 rat anti-mouse CD8 (53−6.7, BioLegend, #100782)
- BODIPY™ C11 581/591 reconstituted to 10 mM in dimethyl sulfoxide (DMSO) (Invitrogen™, D3861)
- Brilliant Stain Buffer Plus (BD Biosciences, #568264)
- Cell samples of interest (in this case, mouse splenocytes)
- Dimethyl sulfoxide (DMSO, Sigma-Aldrich, D8418)
- Dulbecco's phosphate-buffered saline (PBS, Gibco™, #14190144)
- Ethylenediaminetetraacetic acid (EDTA, Sigma-Aldrich®, E5134)
- FC block – purified rat anti-mouse CD16/32 antibody 93, Biolegend, #101302
- Ferrostatin-1 (Fer-1, Sigma-Aldrich®, SML0583)
- Fetal bovine serum (FBS), qualified (Gibco™, #10099141)
- HEPES (1 M, Gibco™, #15630080)
- Minimal essential media non-essential amino acids solution (100X, Gibco™, #11140050)

- Penicillin-Streptomycin-Glutamine (100X, Gibco™, #10378016)
- RPMI 1640 Medium, no glutamine (Gibco™, #21870076)
- Sodium pyruvate 100 mM, Gibco™, #11360070
- Viability dye for flow cytometry, such as ViaDye™ Red (Cytek®, R7-60008)

2.3 Equipment

- Benchtop centrifuge, such as Centrifuge 5810 R (Eppendorf)
- Cell counter, such as Countess™ 3 (Invitrogen™)
- Cell culture incubator, such as CO_2 Incubator (Binder)
- Spectral flow cytometer equipped with at least blue and yellow-green lasers (ideally five-laser configuration), such as Cytek® Aurora (*see Note 1*)

2.4 Software

- FlowJo™ v. 10.9.0 (FlowJo LLC)
- Prism v. 9.5.1 (GraphPad)
- SpectroFlo® v. 3.3.0 (Cytek®)

3. Methods

3.1 Spectral flow cytometer setup

3.1.1 Generation of 'lipid ROS high' and 'lipid ROS low' samples (see Note 2 and 3)

1. Isolate a single cell suspension of the cells of interest using previously defined protocols, such as disaggregating a mouse spleen and performing red blood cell lysis to isolate splenocytes (*see Note 4*).
2. Resuspend the cells in cell culture media, such as complete-RPMI 1640 for T cell culture (*see Note 5*).
3. Perform a cell count and seed an appropriate and consistent number of cells per well in a 96 well round bottom plate (*see Note 6*). For splenocytes, 2×10^6 cells per well is ideal (*see Note 7*).
4. Wash the wells with cell culture media up to a total volume of 200 μL per well.
5. Centrifuge the plate at 400 × *g* for 5 min at 4 °C to pellet the cells, and flick off the supernatant into an appropriate waste container.
6. Resuspend the cells with 200 μL of cell culture media per well, containing a lipid ROS-inducing drug (such as 1 μM RSL3) or a lipid ROS-reducing drug (such as 10 μM Fer-1; *see Note 8 and 9*).
7. Incubate the plate at 37 °C and 5 % CO_2 in an incubator for 2 h.

3.1.2 Initial titration of BODIPY C11 on spectral flow cytometry

1. Seed multiple wells of the previously generated 'lipid ROS high' and lipid ROS 'low' samples in a 96 well round bottom plate, alongside additional untreated cells to use as an unstained control.
2. Centrifuge the plate at 400 × *g* for 5 min at 4 °C and resuspend the wells in 50 μL of BODIPY C11 in PBS (or PBS only for the unstained control). As the initial titration, stain one well each of the 'lipid ROS high' and 'lipid ROS low' samples with 2-fold serial dilutions of BODIPY C11 in PBS, starting from 1 μM BODIPY C11 (1/10 000 dilution of the 10 mM stock) and getting less concentrated.
3. Incubate the plate at 37 °C and 5 % CO2 in an incubator for 30 min (*see Note 10*).
4. Wash by adding 150 μL of PBS to each well and centrifuge the plate at 400x*g* for 5 min at 4 °C.
5. Resuspend in 200 μL of PBS for acquisition at flow cytometry.
6. Acquire the unstained control sample to set appropriate gains for forward scatter (FSC) and side scatter (SSC), and to determine autofluorescence and background spectrum. Use the standard gains and laser settings determined by Cytometer Setup and Tracking (CS&T) for all other parameters.
7. Acquire each of the BODIPY C11-stained samples of 'lipid ROS high' and 'lipid ROS low' cells. Aim to acquire > 50,000 events in P3 (see gating below).
8. For titration gating (Fig. 2A), proceed as follows:
   - **a.** P1 – Gate on the cells of interest by FSC vs SSC
   - **b.** P2 – Gate on single cells by FSC-A vs FSC-H
   - **c.** P3 – Gate on single cells by SSC-A vs SSC-H
9. Using the raw files (not unmixed) view the BODIPY C11 spectra for the 'lipid ROS low' samples at each concentration of BODIPY C11. Identify the highest concentration of BODIPY C11 where the intensity is 'on scale' for all detectors, particularly those on the YG laser (Fig. 2B, *see Note 11*).
10. Export the raw (no unmixing applied).fcs files.

3.1.3 Unmixing of BODIPY C11 on spectral flow cytometry

1. Open an experiment in the SpectroFlo® software containing the unstained control and the BODIPY C11-stained 'lipid ROS high' and 'lipid ROS low'.fcs files from 3.1.2.

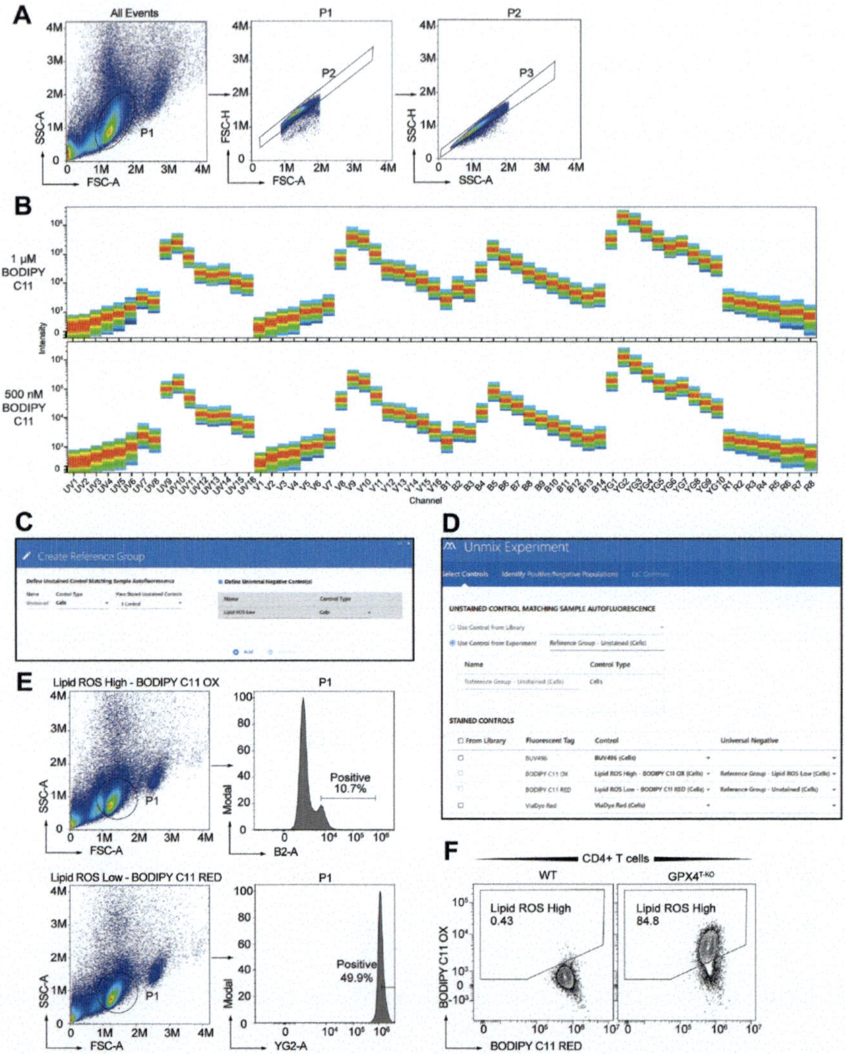

Fig. 2 Experimental setup of BODIPY C11 staining in spectral flow cytometry. A – Representative gating strategy for a titration of BODIPY C11 in lymphocytes. B – Raw BODIPY C11 RED spectra (lipid ROS low samples stained with BODIPY C11 only) at different concentrations of BODIPY C11 in the initial titration. 500 nM is the highest concentration where the full spectrum is on-scale in all channels. C – Example screenshot of adding a universal negative control in SpectroFlo® software. D – Example screenshot of the unmixing wizard in SpectroFlo® showing the selection of reference controls. E – Representative gating of the reference controls to perform unmixing in SpectroFlo®. 'P1' is gated on lymphocytes by forward scatter vs side scatter, then the cells brightest in B2 are gated as the positive for the BODIPY C11 OX reference control, while the cells brightest in YG2 are gated as the positive for the BODIPY C11 RED reference control.

2. Create custom fluorescent tags to represent BODIPY C11 OX (Oxidised) and BODIPY C11 RED (Reduced), respectively. Alternatively, use 'FITC' and 'PE', respectively (*see Note 12*).
3. Under 'Edit Experiment; Groups; Reference Group', select "Define Universal Negative Control(s)" (Fig. 2C).
4. Add a 'universal negative control' and import the.fcs file for the 'lipid ROS low' sample at the highest concentration of BODIPY C11 that is still on scale (identified in 3.1.2 step 9), as this universal negative control.
5. Select 'Unmix', and select the reference controls as follows (Fig. 2D)
   a. BODIPY C11 RED – 'lipid ROS low' sample at the highest concentration of BODIPY C11 that is still on scale (identified in 3.1.2 step 9). Select the 'unstained control' as the universal negative.
   b. BODIPY C11 OX – 'lipid ROS high' sample at the second highest concentration of BODIPY C11 that is still on scale (one dilution down from the BODIPY C11 RED control). Select the 'lipid ROS low' universal negative control (imported in step 4) as the universal negative.
6. Complete the unmixing wizard by selecting the peak channel and gating on the brightest 50 % of the positive population for each control (Fig. 2E, *see Note 13*).
7. Validate the unmixing by plotting BODIPY C11 RED vs BODIPY C11 OX on P3 for the 'lipid ROS high' and 'lipid ROS low' samples. At BODIPY C11 concentrations equal to or lower than those used for unmixing, there should be good resolution of the parameters and there should be a clear increase in BODIPY C11 OX signal in the 'lipid ROS high' samples (Fig. 2F).
8. If possible, identify the optimal concentration of BODIPY C11 (*see Note 14*).

3.1.4 Further titration of BODIPY C11 on spectral flow cytometry – with other markers (Optional, see Note 15)

1. Identify the minimum panel of markers needed to clearly identify the population of interest within the samples (*see Note 16*).

F – Representative spectral flow cytometry data of titrated and unmixed BODIPY C11 staining, gated on viable $CD4^+$ cells from naïve mouse splenocytes. 'WT' represents a wildtype mouse sample as a 'lipid ROS low' sample, while $GPX4^{T\text{-}KO}$ mice (characterised previously (Matsushita et al., 2015)) were used as a 'lipid ROS high' sample to validate BODIPY C11 staining.

2. Generate 'lipid ROS high' and 'lipid ROS low' controls as per 3.1.1., and seed in a 96 well round bottom plate alongside additional untreated cells or beads as reference controls for the other markers in the panel (*See Note 17 and 18*).
3. Perform multi-colour staining of markers alongside BODIPY C11 and staining of reference controls as described subsequently in 3.2.
4. Acquire each sample on the spectral flow cytometer. Aim to acquire > 5000 events in the population(s) of interest. Perform gating as appropriate for the panel and samples.
5. Perform unmixing as per 3.1.3 using fresh or imported reference controls for BODIPY C11, and using standard principles of unmixing for other fluorophores (University of Chicago, 2021).
6. Identify the optimal concentration of BODIPY C11 for the population and cells of interest (*see Note 14*).

3.2 High-parameter spectral flow cytometry using BODIPY C11

3.2.1 Sample preparation – antibody + BODIPY C11 staining (Fig. 3A, see Note 19)

1. Isolate a single cell suspension of the cells of interest using previously defined protocols, such as disaggregating a mouse spleen and performing red blood cell lysis to isolate splenocytes (*see Note 4*).
2. Perform cell count and seed cell samples in a 96 well round bottom plate for staining, using the standard cell number determined in 3.1.1. step 3 (*see Note 7*). Seed additional cells or beads to generate reference controls, plus an additional well for a BODIPY C11 fluorescence minus one (FMO, *see Note 17, 18 and 20*).
3. Wash the cells with PBS up to a total volume of 200–300 μL per well. Centrifuge the plate at 400 × *g* for 5 min at 4 °C.
4. Resuspend the cells with 50 μL of viability dye diluted in PBS (e.g. ViaDye Red, *see Note 21*). Incubate at 4 °C for 20 min in the dark (*see Note 10*).
5. Wash the cells by adding 150 μL of PBS per well and centrifuging at 400 x *g* for 5 min at 4 °C.
6. (Optional, *see Note 22*) Resuspend the cells with 50 μL of FC block. Incubate at 4 °C for 10 min
7. Resuspend the cells with 50 μL of surface antibody cocktail in fluorescence-activated cell sorting (FACS) buffer (*see Note 21, 23 and 24*), including any biotin-conjugated or primary antibodies requiring a two-step stain (e.g. αCD4-Rb IgG). See *Note 25* if step 6 was performed and FC block was used. Incubate at 4 °C for 30 min

8. Wash the cells by adding 150 μL of FACS buffer per well and centrifuging at 400 × *g* for 5 min at 4 °C.
9. Wash the cells by resuspending in 200 μL of PBS per well and centrifuging at 400 × *g* for 5 min at 4 °C.
10. Resuspend the cells with 50 μL of BODIPY C11 diluted in PBS at the previously titrated concentration identified in 3.1.3 step 8 or 3.1.4 step 6. If any other surface antibodies (particularly streptavidin-conjugated fluorophores or secondary antibodies to bind the primary antibodies stained in step 7) need to be stained, add them into the staining cocktail with BODIPY C11 (*see Note 26*). Do not add BODIPY C11 to the FMO well (add only the other fluorophores).
11. Incubate the plate at 37 °C and 5 % CO2 in an incubator for 30 min
12. Wash the cells by adding 150 μL of PBS per well and centrifuging at 400 × *g* for 5 min at 4 °C.
13. Resuspend in 200 μL of PBS for acquisition at flow cytometry (*see Note 27 and 28*).
14. Acquire all reference controls and BODIPY C11-stained samples. Aim to acquire > 5 000 events for the population(s) of interest. Gating will be determined by the panel and experimental question.

3.2.2 BODIPY C11 data analysis

1. Perform unmixing as per 3.1.3 using fresh or imported reference controls for BODIPY C11, and using standard principles of unmixing for other fluorophores (University of Chicago, 2021).
2. Assess the quality of the unmixing by comparing the BODIPY C11-stained samples with the BODIPY C11 FMO. BODIPY C11 may increase the spread of many fluorescent markers, however, it should not significantly alter the percentages of cells in each population after gating (Fig. 3B). If unmixing errors are present, *see Note 29* for troubleshooting.
3. Export the unmixed data and import into analysis software that allows the derivation of new parameters, such as FlowJo™.
4. Select 'Derive Parameters' and create a new parameter called 'BODIPY C11 Ratio'. Enter the formula as: Comp-BODIPY C11 OX ÷ Comp-BODIPY C11 RED. Try different methods of scaling the new parameter's axis, beginning with 'bi-exponential' or 'arcsinh' scaling, and adjust the axis to ensure the data is on an appropriate scale (*see Note 30 and 31*, Fig. 3C).

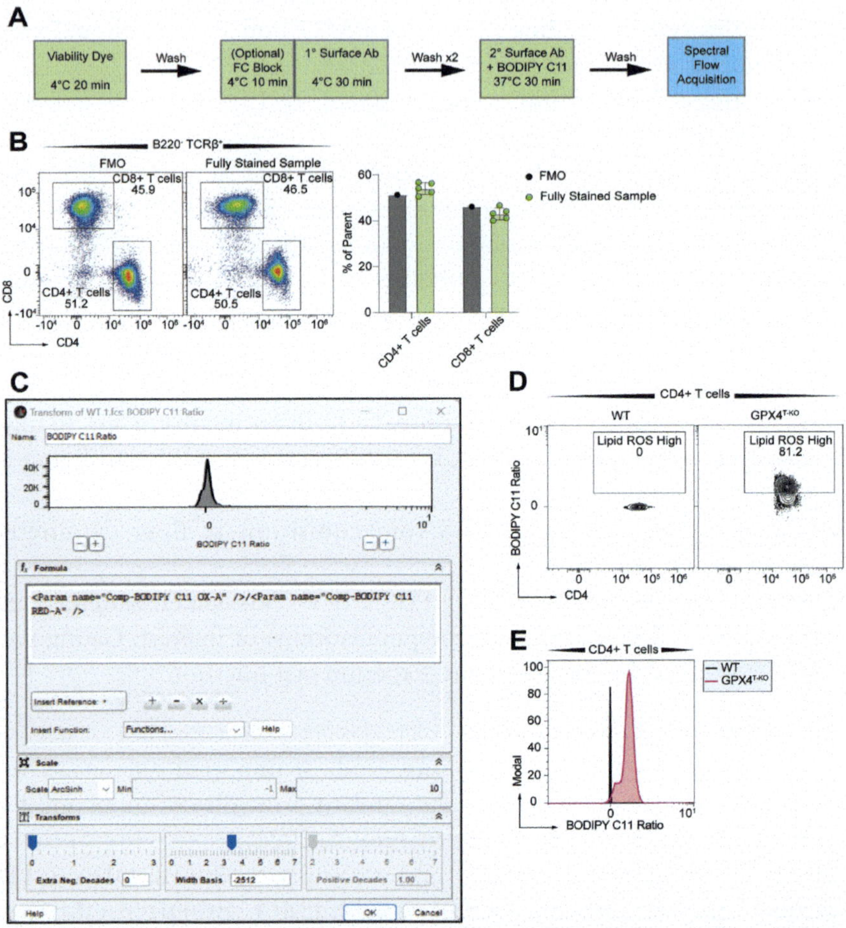

Fig. 3 Staining and analysis of BODIPY C11 in high-parameter spectral flow cytometry. A – Summary schematic showing the staining protocol for BODIPY C11 in a high-parameter spectral flow cytometry panel. B – Example data from a high-parameter panel including BODIPY C11 alongside many other markers, with a comparison of the BODIPY C11 FMO sample and a fully stained sample. Gated on B220^{-} TCRβ^{+}, the CD4 vs CD8 plot confirms that the BODIPY C11 unmixing is accurate, as there are no major differences in the frequency of each population. N = 5 mice for the fully stained sample. The FMO sample was pooled from all five mice. C – Example screenshot from FlowJo™ software showing the 'derive parameters' window with the suggested settings for calculating the BODIPY C11 ratio. BODIPY C11 ratio can be calculated as 'BODIPY C11 OX ÷ BODIPY C11 RED', and is usually best scaled using an 'arcsinh' function, with representative transformation values shown in the screenshot. D-E – Representative data using the derived 'BODIPY C11 Ratio' parameter to gate on lipid ROS high cells (D) or compare BODIPY C11 Ratio MFI using histograms (E). WT and GPX4$^{T\text{-}KO}$ samples were as described in Fig. 2F, and were gated on CD4+ T cells. 'FMO' = Fluorescence minus one.

5. Perform gating as appropriate for the panel and experimental questions. Use a BODIPY C11 OX vs BODIPY C11 RED plot (or a plot involving BODIPY C11 Ratio) to gate on lipid ROS high vs lipid ROS low populations, or calculate the mean fluorescence index (MFI) of BODIPY C11 Ratio for the population(s) of interest (Fig. 3D and E).

4. Notes

1. Our optimization and validation experiments have all been performed using the Cytek® Aurora and SpectroFlo® software, however, all aspects of the protocol should be compatible and easy to adapt to another spectral flow cytometer or other data acquisition software. Similarly, all details of the protocol were written to suit the Aurora with a five-laser configuration, but can be adapted to a different configuration as well.
2. The simplest method to achieve this will be dependent on the cell and sample type used. In mouse T cells, the provided protocol on in vitro RSL3 vs Fer-1 treatment was the simplest method to generate high lipid ROS vs low lipid ROS controls. However, untreated splenocytes from the previously characterised $GPX4^{T\text{-}KO}$ mice (Matsushita et al., 2015) were also used as lipid ROS high samples for the validation and titration of BODIPY C11 staining (Figs. 2 and 3), and certainly any convenient strategy to induce lipid ROS can be used for this protocol.
3. As per the guidelines for spectral flow cytometry reference controls (University of Chicago, 2021), the 'lipid ROS high' sample should have the same or more lipid ROS than the experimental samples (to ensure the BODIPY C11 staining is as bright or brighter than the multi-colour sample). Similarly, the 'lipid ROS low' sample should have the same or less lipid ROS than the experimental samples.
4. During sample preparation, aim for consistency between all samples in terms of cell viability and presence of debris as these factors may impact on BODIPY C11 staining, particularly in terms of staining intensity.
5. Complete-RPMI 1640 can be prepared as follows: RPMI 1640 (without L-Glutamine, with phenol red) + 10 % heat-inactivated fetal bovine serum, 1 % Penicillin-Streptomycin-Glutamine, 10 mM HEPES, 1X minimal essential media non-essential amino acids, 1 mM sodium pyruvate and 55 μM 2-mercaptoethanol.

6. All volumes and steps in this protocol are optimized for a 96 well round bottom plate, however any appropriate vessel and staining volumes used for flow cytometry would be suitable.
7. Any cell number appropriate for flow cytometry is acceptable, however, the cell number should be kept consistent for all reference controls, titrations and experimental samples, as BODIPY C11 staining intensity is cell number dependent. Further, as debris and dead cells can affect BODIPY C11 staining (discussed in *Note 4*), aim to seed a consistent number of total cells (live + dead cells) per well, rather than a consistent number of viable cells.
8. Select an appropriate pair of drugs/reagents to induce and reduce (respectively) lipid ROS in the cells of interest, as different cell types will have varied sensitivity to each ferroptosis-modulating drug. Titration of drug type, concentration and incubation times may be required to induce high lipid ROS without inducing ferroptotic cell death.
9. Avoid concentrations above 5 μM for RSL3, as it may have non-ferroptosis-related side effects at higher concentrations (Park et al., 2023).
10. Once staining with fluorescent markers has begun (including BODIPY C11), keep samples in the dark wherever possible to avoid photobleaching.
11. BODIPY C11 is a very bright dye and the staining is often brighter than the maximum intensity that a flow cytometer can measure. For spectral flow cytometry, the full spectrum of each dye should be on scale in order to accurately unmix the samples.
12. The naming and details of the 'fluorescent tags' section is purely for appearance and communication of the data, and is ultimately arbitrary. If creating custom fluorescent tags, BODIPY C11 OX should have B2 as its peak channel, and BODIPY C11 RED can have YG1 or YG2 as its peak channel.
13. B2 must be used as the peak channel for BODIPY C11 OX. YG2 is preferred for BODIPY C11 RED (as it is brightest in this channel), but YG1 can be used interchangeably, especially if the YG2 intensity is close to or partially at the maximum intensity.
14. The optimal concentration of BODIPY C11 is the lowest concentration that shows a clear difference between 'lipid ROS high' and 'lipid ROS low' samples. Titrating down the BODIPY C11 brightness is particularly essential to reduce spreading error when using high-parameter panels in subsequent staining. Usually, the sensitivity and

separation between high and low lipid ROS decreases as the concentration of BODIPY C11 decreases, so the optimal concentration represents the 'sweet spot' between these two factors.

15. Optional if the optimal concentration could be identified in 3.1.3 step 8. This is only the case if the cells of interest can be accurately identified by FSC and SSC only as in the initial titration (e.g. if the samples were a purified/homogenous cell culture), otherwise a second titration with other fluorescent markers is essential.

16. In panel design, account for BODIPY C11 spillover where possible by analysing the raw spectrum of BODIPY C11 and avoiding fluorophores with peak channels that overlap with the regions where BODIPY C11 is brightest. Fluorophores with a peak channel of B2, YG1 or YG2 are not compatible with BODIPY C11. Fluorophores with peak channels of UV9-11, V8-11, B5-7 or YG3-8 should be avoided wherever possible. If necessary to use these channels, prioritise assigning highly expressed (primary staining pattern) and non-critical markers to these fluorophores.

17. As per the guidelines for spectral flow cytometry reference controls (University of Chicago, 2021), references should be single-colour controls stained in the same manner as the multi-colour samples. Best practice is to stain the reference controls alongside the multi-colour samples (with all the same incubations, treatments and washes), but only add a single fluorophore for each reference.

18. For viability dye reference controls, take an aliquot of cells and induce cell death – simply heating to 60 °C for 10 min is usually sufficient. Spike these dead cells into the viability dye reference control well prior to staining to ensure there is a bright positive signal.

19. The following protocol represents the standard staining protocol for flow cytometry in our group, adapted to be suitable for BODIPY C11 staining. A different flow cytometry staining protocol could be adapted to include BODIPY C11 similarly, as long as the following details are kept the same as the provided protocol:

a. BODIPY C11 should be stained in the final step of staining

b. BODIPY C11 should be stained at 37 °C

c. Samples should not be fixed (discussed further in *Note 27*)

20. A BODIPY C11 FMO is a sample stained with all fluorophores in the panel except for BODIPY C11. This allows validation and troubleshooting of the BODIPY C11 unmixing and whether it affects the other markers in the panel.

21. All viability dyes and fluorescent markers should be separately titrated and the saturating antibody dilution used for optimal staining (University of Chicago, 2021).
22. FC block is crucial for staining cells with high non-specific binding, such as B cells. It is recommended but not essential for staining most T cell populations.
23. FACS buffer can be prepared as follows: PBS + 2 % heat-inactivated FBS + 2 mM EDTA.
24. For complex panels (>10 fluorophores), it is best to stain the surface antibodies prior to staining BODIPY C11 for a couple reasons. Firstly, they can be stained at 4 °C in FACS buffer, whereas BODIPY C11 needs to be stained at 37 °C in PBS. Secondly, brilliant stain buffer can be used when staining surface markers prior to BODIPY C11 (*see Note 26*).

    If using FC block, then double the usual antibody concentration to account for the 1:1 dilution with FC block:surface antibody cocktail (e.g. a marker normally stained at 1/200 would need to be 1/100 in the surface antibody cocktail).
25. In our hands, BODIPY C11 staining works best in PBS without any other buffers (e.g. brilliant stain buffer). If an antibody requires a specific buffer other than PBS, it should be stained prior to BODIPY C11 staining, such as in 3.2.1 step 7. Keep the samples in PBS only after staining with BODIPY C11.
26. BODIPY C11 should only be used in live, unfixed cells for FACS. Fixation (for intracellular or intranuclear antibody staining) dramatically reduces the intensity and sensitivity of the dye, meaning that only surface markers should be stained in a panel with BODIPY C11.
27. After staining with BODIPY C11, keep samples on ice and in the dark to maintain sample and fluorophore quality. Proceed to sample acquisition as quickly as possible since samples are unfixed and will degrade over time.
28. If BODIPY C11 unmixing errors are present, take the following steps in order until the problem is solved:
    a. Ensure BODIPY C11 reference controls are brighter than the BODIPY C11 fluorescence in the multi-colour samples, but all are still 'on-scale'. That includes both the YG1/2 intensity for the BODIPY C11 RED reference control, and the B2 intensity for the BODIPY C11 OX reference control.

b. Titrate BODIPY C11 down further to a lower concentration. Be aware that the sensitivity may decrease. If this is a problem, consider keeping the original BODIPY C11 concentration and proceeding to 'c'.

c. Adjust the fluorescent markers included in the panel to avoid problematic channels.

d. If all else fails, consider applying a manual compensation matrix on top of the unmixing algorithm (easily facilitated by SpectroFlo® or FlowJo™ software). Avoid making large changes, and use only the reference controls to adjust the compensation values and correct unmixing errors (University of Chicago, 2021).

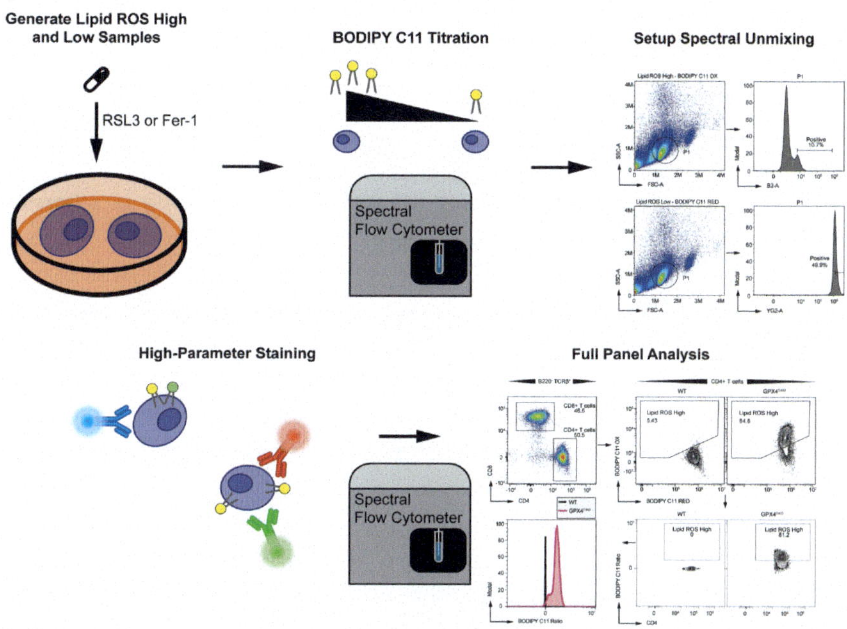

Fig. 4 Schematic illustration of the overall protocol - Lipid ROS Measurement by Spectral Flow Cytometry based on the ratiometric dye BODIPY C11. Firstly, cells are treated to generate 'lipid ROS high' and 'lipid ROS low' controls, likely using drugs such as RSL3 and Fer-1. These samples are used to perform a titration of BODIPY C11, analysed on a spectral flow cytometer, and the spectral unmixing algorithm can be set up. High-parameter staining is then performed using various fluorescent antibodies alongside BODIPY C11 to stain cells, and data is acquired on the spectral flow cytometer. The full panel of unmixed data can then be analysed to detect lipid ROS in individual cell subsets, such as CD4+ T cells as shown in this representative data. Representative data from Figs. 2E-F, 3B and 3D-E are included here. See the individual figures for the description and source of this data.

Choosing a scaling method often requires some optimization and can vary depending on the distribution of the data in question. As BODIPY C11 Ratio data can produce negative values (if the BODIPY C11 OX MFI intensity is negative), avoid 'logarithmic' scaling. Bi-exponential or arcsinh are both equally valid for ratiometric data, so whichever gives the best resolution should be used. Use the previously generated 'lipid ROS high' and 'lipid ROS low' controls to determine which method provides the optimal scaling.

While a simpler approach exists to calculate the BODIPY C11 Ratio for a group/sample by dividing the BODIPY C11 OX MFI of the group/sample by its MFI for BODIPY C11 RED, deriving a new parameter for the BODIPY C11 Ratio is a more accurate method as it considers the individual lipid ROS measurement for each cell/event.

5. Concluding remarks

Lipid ROS detection through BODIPY C11 staining is a cornerstone of modern ferroptosis research, however, existing literature fails to acknowledge the significant challenges associated with using this dye for flow cytometry. Notably, the broad spillover and ratiometric properties of the dye have limited research groups to using only traditional flow cytometry with small panels (no more than ten fluorophores), where the compensation can be manually or artificially calculated using SCCs that are not 100% accurate to the BODIPY C11 OX spectrum (e.g. using a generic FITC antibody as a surrogate).

In this protocol (summarised in Fig. 4), we provide a novel strategy for unmixing BODIPY C11 as a ratiometric dye in spectral flow cytometry, including suggested methods to generate SCCs via lipid ROS low and lipid ROS high samples, an unmixing method to separate the BODIPY C11 OX spectrum from the BODIPY C11 RED background, and solutions for the other challenges of BODIPY C11 such as its extensive spillover and extremely bright fluorescence. In our testing, this protocol has enabled the use of BODIPY C11 in spectral flow cytometry panels of over 30 fluorophores, allowing far deeper exploratory profiling of lipid ROS in different cell types simultaneously.

The unmixing strategy was developed following inspiration from previous work on adapting Forster resonance energy transfer (FRET) imaging for spectral flow cytometry (Henderson et al., 2022), whereby the authors utilised 'FRET-High' and 'FRET-Low' controls to generate the required reference spectra. Similarly, we employed 'lipid ROS high' and 'lipid ROS low' samples in our protocol, but extended beyond this with the additional strategy to use the 'lipid ROS low' sample as the 'universal negative' to unmix BODIPY C11 OX. This novel unmixing strategy effectively separated the BODIPY C11 OX and BODIPY C11 RED signatures, as the unmixing algorithm treated the BODIPY C11 RED spectrum as 'background' in the 'lipid ROS high' sample, meaning that the specific spectral differences between BODIPY C11 staining in the lipid ROS high vs lipid ROS low sample were recorded as the BODIPY C11 OX reference spectrum.

As such, this method was roughly five to eight times more sensitive than the equivalent performed using traditional flow cytometry and standard compensation (data not shown), which relies solely on the fluorescence in a single channel to measure BODIPY C11 OX. Therefore, this method represents a significant improvement on measuring lipid ROS compared to traditional flow cytometry, uneven for very small panels. Although adapting BODIPY C11 staining to spectral flow requires some titration and optimization (detailed in 3.1), whereas BODIPY C11 requires much less set up in traditional flow cytometry, the usefulness and high sensitivity of this method for ferroptosis research in broad cell types (such as immune cells) opens broad new avenues. In our testing, BODIPY C11 could be added to most spectral flow cytometry panels with minimal issues, optimized by adjusting the fluorophores for a few markers only. However, the most optimal staining can be achieved by designing a spectral flow cytometry panel to account for BODIPY C11 from scratch, as per the considerations in *Note 16*.

Despite its sensitivity, a limitation on the use of BODIPY C11 for lipid ROS detection is that the dye functions much less effectively in fixed samples, as per the discussion in *Note 27*. This limits the method to detecting lipid ROS in panels and cell types that can be identified by surface staining only, as fixation is generally required to stain intracellular or intranuclear markers by flow cytometry. Fluorescent reporters (such as BFP-tagged proteins) could be used to bypass this limitation, however the most popular reporters, like GFP and mCherry, have fluorescent spectra that are highly similar to BODIPY C11 OX and BODIPY C11 RED respectively, meaning they should also be avoided. Moreover, as with all flow cytometry, this method is more challenging to perform when analysing rarer cell populations. Additional considerations should be undertaken for detecting lipid ROS in these rare cell populations, such as staining and analysing a larger number of total cells (to ensure sufficient events are recorded for the cells of interest), and carefully validating the unmixing (to ensure these rare events are not mistakenly missed).

Outside of the direct applications of this protocol to facilitate exploratory ferroptosis research, the methodology for unmixing BODIPY C11 as a ratiometric dye is likely generalizable to other ratiometric dyes. The general rationale of generating samples that are high in each parameter measured by the ratiometric dye has now been validated by our method and the previous work in FRET analysis (Henderson et al., 2022). Moreover, the strategy of using one of the reference controls as the 'universal negative' for the other reference control can be translated to any other ratiometric dye where complete fluorescent conversion to the alternate spectrum of the dye cannot be achieved, as in the case of BODIPY C11. Hence, the protocol herein represents a useful technique for both ferroptosis and other research fields, harnessing expanded capabilities of spectral flow cytometry beyond standard antibody staining.

Acknowledgments

The authors acknowledge the Translational Research Institute (TRI), the Biological Research Facility (BRF) and the Flow Cytometry Core Facility (Lucie Leveque-El Mouttie, Dalia Khalil, Yitian Ding, Andy Wu and Jintao Guo) for providing the excellent research environment and core facilities that enabled this research. In particular, thanks to Dr Leveque-El Mouttie for her help in troubleshooting the initial use of BODIPY C11 on the Cytek® Aurora. Additionally, we would like to thank the TRI Flow Cytometry Core Facility for their generous support through the High-Parameter Panel Development Reagent Support Fund. D.Y. is supported by Australian National Health and Medical Research Council (NHMRC) grants (GNT2009554), Medical Research Futures Fund (MRFF) National Critical Infrastructure Initiative (NCRI000155), and fundings from Children's Hospital Foundation, the Australia and New Zealand Society for Immunology (AbbVie New Horizon Award) and Tour de Cure (RFP-660–2023, RSP-585–2024 and RSP-616–2024).

Competing interests

The authors have no competing interests to declare.

References

Astakhova, E. A., Gubaeva, A. S., Naumova, D. A., Egorova, A. E., Maznina, A. A., Rybkina, I. G., ... Volchkov, P. Y. (2025). Spectral flow cytometry: The current state and future of the technology. *International Journal of Molecular Sciences, 26*(12), https://doi.org/10.3390/ijms26125911.

Dixon, S. J., & Pratt, D. A. (2023). Ferroptosis: A flexible constellation of related biochemical mechanisms. *Molecular Cell, 83*(7), 1030–1042. https://doi.org/10.1016/j.molcel.2023.03.005.

Dojindo (2023). *Lipid peroxidation probe-BDP 581/591 C11-*. Retrieved 06/12/2023 from https://www.dojindo.com/ASIA/products/L267/.

Doll, S., Proneth, B., Tyurina, Y. Y., Panzilius, E., Kobayashi, S., Ingold, I., ... Conrad, M. (2017). ACSL4 dictates ferroptosis sensitivity by shaping cellular lipid composition. *Nature Chemical Biology, 13*(1), 91–98. https://doi.org/10.1038/nchembio.2239.

Henderson, J., Havranek, O., Ma, M. C. J., Herman, V., Kupcova, K., Chrbolkova, T., ... Davis, R. E. (2022). Detecting Förster resonance energy transfer in living cells by conventional and spectral flow cytometry. *Cytometry Part A, 101*(10), 818–834. https://doi.org/10.1002/cyto.a.24472.

Jin, X., Tang, J., Qiu, X., Nie, X., Ou, S., Wu, G., Zhang, R., & Zhu, J. (2024). Ferroptosis: Emerging mechanisms, biological function, and therapeutic potential in cancer and inflammation. *Cell Death Discovery, 10*(1), 45. https://doi.org/10.1038/s41420-024-01825-7.

Liu, J., Dong, R., Yuan, B., Xie, Y., Feng, Z., Zhou, S., ... Chen, Y. (2025). Immune cells dying from ferroptosis: Mechanisms and therapeutic opportunities. *Cell Death & Disease, 16*(1), 878. https://doi.org/10.1038/s41419-025-08204-9.

Ma, X., Xiao, L., Liu, L., Ye, L., Su, P., Bi, E., ... Yi, Q. (2021). CD36-mediated ferroptosis dampens intratumoral CD8(+) T cell effector function and impairs their antitumor ability. *Cell Metabolism, 33*(5), 1001–1012.e1005. https://doi.org/10.1016/j.cmet.2021.02.015.

Matsushita, M., Freigang, S., Schneider, C., Conrad, M., Bornkamm, G. W., & Kopf, M. (2015). T cell lipid peroxidation induces ferroptosis and prevents immunity to infection. *The Journal of Experimental Medicine, 212*(4), 555–568. https://doi.org/10.1084/jem.20140857.

Mishima, E., Nakamura, T., Doll, S., Proneth, B., Fedorova, M., Pratt, D. A., ... Conrad, M. (2025). Recommendations for robust and reproducible research on ferroptosis. *Nature Reviews. Molecular Cell Biology, 26*(8), 615–630. https://doi.org/10.1038/s41580-025-00843-2.

Morgan, P. K., Pernes, G., Huynh, K., Giles, C., Paul, S., Smith, A. A. T., ... Murphy, A. J. (2024). A lipid atlas of human and mouse immune cells provides insights into ferroptosis susceptibility. *Nature Cell Biology, 26*(4), 645–659. https://doi.org/10.1038/s41556-024-01377-z.

Park, S. Y., Jeong, K. J., Poire, A., Zhang, D., Tsang, Y. H., Blucher, A. S., & Mills, G. B. (2023). Irreversible HER2 inhibitors overcome resistance to the RSL3 ferroptosis inducer in non-HER2 amplified luminal breast cancer. *Cell Death & Disease, 14*(8), 532. https://doi.org/10.1038/s41419-023-06042-1.

Stockwell, B. R. (2022). Ferroptosis turns 10: Emerging mechanisms, physiological functions, and therapeutic applications. *Cell, 185*(14), 2401–2421. https://doi.org/10.1016/j.cell.2022.06.003.

University of Chicago. (2021). Cytek Aurora training course [Video series].

Wiernicki, B., Dubois, H., Tyurina, Y. Y., Hassannia, B., Bayir, H., Kagan, V. E., ... Vanden Berghe, T. (2020). Excessive phospholipid peroxidation distinguishes ferroptosis from other cell death modes including pyroptosis. *Cell Death & Disease, 11*(10), 922. https://doi.org/10.1038/s41419-020-03118-0.

Xu, C., Sun, S., Johnson, T., Qi, R., Zhang, S., Zhang, J., & Yang, K. (2021). The glutathione peroxidase Gpx4 prevents lipid peroxidation and ferroptosis to sustain Treg cell activation and suppression of antitumor immunity. *Cell Reports, 35*(11), 109235. https://doi.org/10.1016/j.celrep.2021.109235.

Xu, S., Chaudhary, O., Rodríguez-Morales, P., Sun, X., Chen, D., Zappasodi, R., ... Kaech, S. M. (2021). Uptake of oxidized lipids by the scavenger receptor CD36 promotes lipid peroxidation and dysfunction in CD8(+) T cells in tumors. *Immunity, 54*(7), 1561–1577.e1567. https://doi.org/10.1016/j.immuni.2021.05.003.

Yang, W. S., SriRamaratnam, R., Welsch, M. E., Shimada, K., Skouta, R., Viswanathan, V. S., ... Stockwell, B. R. (2014). Regulation of ferroptotic cancer cell death by GPX4. *Cell, 156*(1-2), 317–331. https://doi.org/10.1016/j.cell.2013.12.010.

Yang, W. S., & Stockwell, B. R. (2008). Synthetic lethal screening identifies compounds activating iron-dependent, nonapoptotic cell death in oncogenic-RAS-harboring cancer cells. *Chemistry & Biology, 15*(3), 234–245. https://doi.org/10.1016/j.chembiol.2008.02.010.

Yao, Y., Chen, Z., Zhang, H., Chen, C., Zeng, M., Yunis, J., ... Yu, D. (2021). Selenium–GPX4 axis protects follicular helper T cells from ferroptosis. *Nature Immunology, 22*(9), 1127–1139. https://doi.org/10.1038/s41590-021-00996-0.

CHAPTER FIVE

CRISPR-Cas9 screen to identify genes regulating cell death

Zilong Zheng[a] and Daichao Xu[a,b,c,d,*]
[a]Interdisciplinary Research Center on Biology and Chemistry, Shanghai Institute of Organic Chemistry, Chinese Academy of Sciences, Shanghai, P.R. China
[b]State Key Laboratory of Chemical Biology, Shanghai Institute of Organic Chemistry, Chinese Academy of Sciences, Shanghai, P.R. China
[c]Shanghai Key Laboratory of Aging Studies, Shanghai, P.R. China
[d]Shanghai Academy of Natural Sciences (SANS), Fudan University, Shanghai, P.R. China
*Corresponding author. e-mail address: xudaichao@sioc.ac.cn

Contents

Abstract

Regulated cell death mediated by dedicated molecular machines, known as programmed cell death, plays important roles in health and disease. Understanding the mechanisms of cell death is crucial for elucidating the control of cellular homeostasis and developing therapies for related diseases. Despite extensive research efforts spanning decades, many aspects of cell death mechanisms remain elusive, highlighting the need for continued exploration. Here, we describe how to identify novel regulators involved in cell death pathways using a genome-wide screening approach.

Methods in Cell Biology, Volume 210
ISSN 0091-679X, https://doi.org/10.1016/bs.mcb.2026.05.010

1. Introduction

CRISPR (Clustered Regularly Interspaced Short Palindromic Repeats) represents a microbial nuclease mechanism which encompasses a suite of CRISPR-associated (Cas) genes alongside non-coding RNA components that have the capacity to direct the precision of nucleic acid cleavage (Barrangou & Doudna, 2016; Garneau et al., 2010; Wang et al., 2016). Cas9 functions as an RNA-guided endonuclease. Its sequence specificity is primarily determined by Watson-Crick base pairing between the guide RNA and the target DNA sequence, complemented by a direct interaction between Cas9 and a short DNA sequence known as the protospacer-adjacent motif (PAM) (Garneau et al., 2010; Gasiunas et al., 2012; Jinek et al., 2012; Marraffini & Sontheimer, 2008). This unique mechanism allows Cas9 to be reprogrammed to target different DNA sequences simply by altering the sequence of its guide RNA. Thus, Cas9 serves as a versatile and efficient tool for high-throughput, sequence-specific genome editing.

Genome-wide CRISPR knockout screening provides several unique advantages for identifying regulators of cell death. First, the approach enables unbiased and systematic interrogation of the entire protein-coding genome, allowing discovery of unexpected or previously uncharacterized factors that would not emerge from hypothesis-driven studies. Second, CRISPR-mediated gene disruption produces highly penetrant and stable loss-of-function phenotypes, which is critical for studying cell death pathways where partial knockdown (e.g., with RNAi) often leads to ambiguous or incomplete effects. Importantly, this methodology exhibits high versatility, as it can be readily adapted for both positive-selection screens, where resistance-conferring sgRNAs are enriched, and negative-selection screens, in which essential regulators of cell death are depleted under selective conditions. In addition, the approach is compatible with diverse cell systems, including primary cells and organoids, and supports a wide range of biological perturbations.

Here we describe how to use GeCKO v2 libraries (Shalem et al., 2014, Addgene: #1000000053) to identify key genes regulating cell death (Fig. 1). The GeCKO v2 libraries are designed to facilitate gene knockout studies in the human and mouse genomes. Each species-specific library is divided into two half-libraries, designated as A (3 sgRNAs per gene) and B (3 sgRNAs per gene). Together, these half-libraries provide a total of six single guide RNA (sgRNA) sequences per gene, with three sgRNAs contained in each half-library. For optimal results, it is recommended to

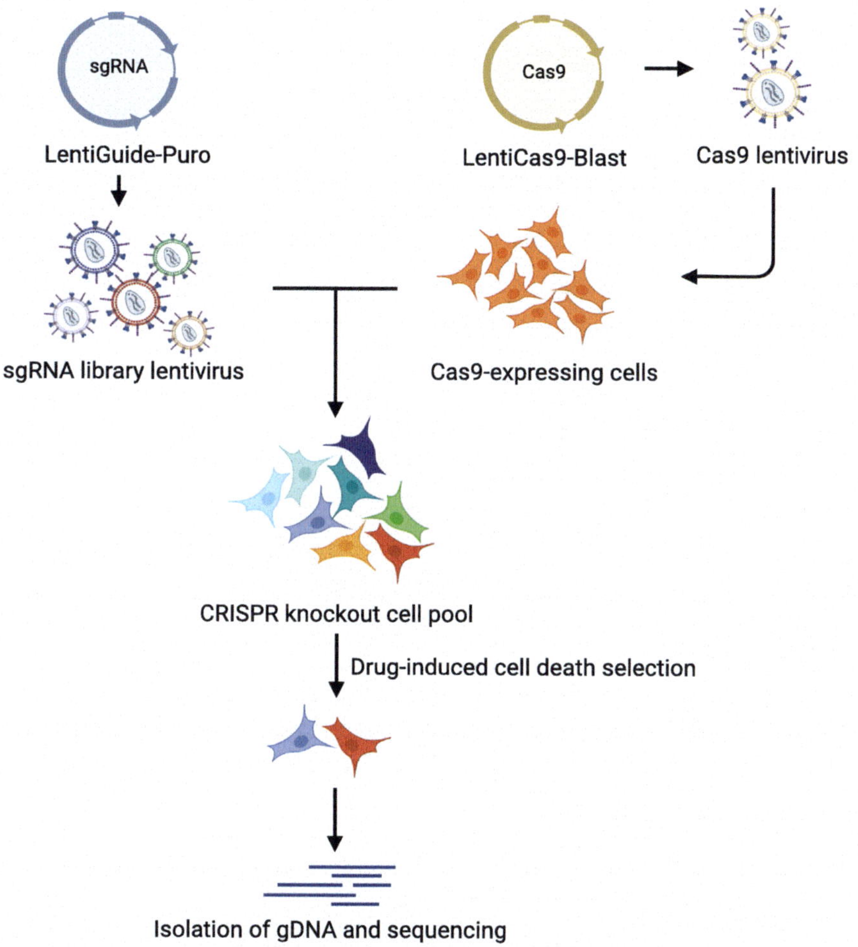

Fig. 1 Overview of the genome-wide CRISPR screening workflow. Schematic illustration of the experimental pipeline, including lentiviral sgRNA library transduction, generation of Cas9-expressing cells, drug treatment, genomic DNA extraction, and next-generation sequencing.

screen the complete library (both A and B) whenever feasible. However, if achieving adequate coverage with the full library is not possible, screening can be conducted using a single half-library. This approach ensures flexibility while maintaining the robustness of the gene knockout screening process, for example, by the use of partial libraries, and compatibility with both positive and negative selection screens.The GeCKO v2 libraries is provided in 2 vector (lentiCas9-Blast and lentiGuide-Puro) format. The two-vector system, which utilizes the lentiGuide-Puro vector to deliver the sgRNA library, offers the advantage of achieving higher viral titers for library transduction. However, this system requires that the target cells already express Cas9, typically achieved through stable genomic integration using the lentiCas9-Blast vector. This two-vector design supports efficient gene editing while providing flexibility for experimental workflows, as Cas9 can be stably introduced once and reused across multiple screens, and the sgRNA library can be delivered independently, allowing researchers to package the sgRNA library separately and achieve higher viral titers for transduction.

2. Materials

2.1 Common disposables

- Cell culture flasks 60 mm (#353802, Corning)
- Cell culture flasks 100 mm (#353803, Corning)
- 24.5 cm^2 bioassay plates (#240835, Thermo Fisher Scientific)
- Corning® 96-well Clear Flat Bottom Polystyrene TC-treated Microplates (#3598, Corning)
- Falcon® 15 mL High Clarity PP Centrifuge Tube (#352096, Corning)
- Falcon® 50 mL High Clarity PP Centrifuge Tube, Conical Bottom, with Dome Seal Screw Cap, Sterile (#352070, Corning)

2.2 Cells and reagents

- lentiCas9-Blast (Addgene, cat. no. 52962),
- lentiGuide-Puro (Addgene, cat. no. 52963),
- Mouse GeCKO v2 libraries (Addgene, cat. no. 1000000053),
- pMD2. G (Addgene, cat. no. 12259),
- psPAX2 (Addgene, cat. no. 12260),
- ElectroMAX™ DH10B (#18290015, Invitrogen)
- Blasticidin S HCl (#1113903, Thermo Fisher Scientific),
- Puromycin dihydrochloride (#A1113803, Thermo Fisher Scientific),
- QIAquick PCR Purification Kit (#28104, Qiagen)
- QIAquick Gel Extraction Kit (#28704, Qiagen),
- ZYMO Quick-DNA Midiprep Plus kit (#D4075, Zymo research),
- CellTiter-Glo Luminescent Cell Viability Assay (#G7571, Promega)
- Qubit dsDNA HS Assay Kit (#Q32851, Thermo Fisher Scientific)
- S-1.2x-TN5 Mix (Tsingke TDE0505)
- MiSeq Reagent Micro Kit V2 (Illumina MS-102-2002)

3. Methods

3.1 Amplification of pooled sgrna library

3.1.1 Electroporation of library plasmid

1. Add 100 ng of the library plasmid to 25 μL of electrocompetent cells (e.g., ElectroMAX™ DH10B) with a transformation efficiency of ≥ 10^9 cfu/μg. Perform electroporation according to the recommended parameters of the electroporation instrument.
2. After electroporation, add 975 μL of SOC medium (2 % (w/v) tryptone, 0.5 % (w/v) yeast extract, 10 mM NaCl, 2.5 mM KCl, 10 mM $MgCl_2$, 10 mM MgSO4, 20 mM Glucose), mix well, and transfer to a culture tube. Add 1 mL of recovery medium to the tube and mix again. Repeat this process three times to obtain a total of 4 tubes of electroporation products.
3. Place the tubes in a shaker and incubate at 250 rpm and 37 °C for 1 h.

3.1.2 Amplification of library culture and calculation of transformation efficiency

1. Mix the 4 tubes of electroporation products together (8 mL in total), take 10 μL, and dilute it with 990 μL of SOC medium. Plate 20 μL of the diluted solution onto a 10 cm bacterial culture dish and incubate at 37 °C overnight.
2. Count the number of colonies on the plate. The total number of bacterial colonies is calculated as the number of colonies per plate multiplied by 40,000. If the total colony number exceeds 3×10^6, this indicates that each sgRNA in the library is represented by at least 50 independent clones, and the electroporation efficiency meets the required standard. According to the above calculation, if the number of colonies on each plate exceeds 75, the electroporation efficiency is considered sufficient.
3. Plate 4 mL electroporation products of transformation on each of 2 pre-warmed 24.5 cm^2 bioassay plates (100 μg/mL ampicillin) using a spreader. Alternatively, plate 400 μL per 10 cm bacterial culture dish (100 μg/mL ampicillin), totaling about 20 dishes, and incubate at 37 °C overnight.

3.1.3 Collection of transformation products and plasmid extraction

1. Add 10 mL LB to each dish 24.5 cm^2 bioassay plates, use a scraper to collect the colonies, and transfer the bacterial suspension to a 50 mL

centrifuge tube. For 10 cm bacterial culture dish, add 500 μL of LB medium to each dish to collect colonies.

2. Repeat the above steps until all the colonies are collected.
3. After centrifugation at 4000 *g* for 15 min at room temperature, discard the supernatant and weigh the pellet (bacteria).
4. Extract the plasmid according to the instructions of the large-scale plasmid extraction kit.

3.2 Lentivirus production and titer

3.2.1 Virus packaging

Mix the library plasmid (lentiGuide-Puro) or lentiCas9-Blast with virus packaging plasmids (pMD2.G and psPAX2) and co-transfect them into 293 T cells. Cells in 125 cm dishes are optimal for transfection at 80–90 % confluency. Transfection method could use Lipo2000, polyethylenimine (PEI) or other transfection reagent. Collect the virus 48 or 72 h later, concentrate it, and store it at −80°C for future use.

3.2.2 Infectious MOI determination

Dilute the library virus into different gradients, such as MOI = 0.3, 0.5, 1, 5, 10, 30, 100, and infect the target cells (cell confluency of 30–50 %). Set up 6 wells (96-well plate) for each gradient. After 48 h of infection, add appropriate concentration of puromycin (see Note 5). Stop the drug screening until all cells in the blank group (uninfected cells) are dead. Quantify the cell viability for each condition using CellTiter Glo according to the manufacturer's protocol. For each virus condition, calculate the MOI as the average luminescence, or viability, of the 3 wells in the condition with antibiotic selection divided by the average luminescence of the 3 wells in the condition without antibiotic selection. A linear relationship between lentivirus supernatant volume and MOI is expected at lower volumes, with saturation achieved at higher volumes.

3.3 Lentiviral transduction, screening and suquencing

3.3.1 Generation of a cell line with stably expressed Cas9 components

1. Transduce the relevant cell line with virus which are produced by HEK293T transfected with lentiCas9-Blast, pMD2.G and psPAX2 at an MOI of 0.3.
2. 24 h after lentiviral transduction, replace the medium with DMEM containing the blasticidin.

3. Select Cas9 stable expressed cells for other 5−7 days, until there are no viable cells in the no-virus control
4. After selecting for successfully transduced cells, allow the cells to recover from selection by culturing in normal medium for 2−7 d before transducing with the sgRNA library.

3.3.2 Library virus infection and screening

1. Confirm cell and virus quantities. To ensure that most cells receive only one genetic perturbation, transduce the sgRNA library at an MOI = 0.3.
2. Cell count = (gRNA number × gRNA coverage) / 30 %

   gRNA coverage > 500

   Virus amount = Cell count × infectious MOI
3. Amplify the Cas9 stable expressed cells according to the calculated cell count and prepare an adequate amount of virus.
4. Infect the target cells with the library virus. 24−48 h after viral infection, add an appropriate concentration of puromycin and screen for 5–7 days.
5. After puromycin screening, divide the cells into experimental and control groups. The number of the control cells (total sgRNA library) is at least 3×10^7.
6. Add the target drug to the experimental group for screening. For example, treatment with methyl-beta-cyclodextrin (MCD) induces acute cholesterol depletion, which leads to cell death. Prepare DMEM medium supplemented with 2 % fetal bovine serum (FBS, lipid-free) and 5 mM MCD. Remove the existing medium from the cells in the treatment group, wash three times with PBS, and replace with the MCD-containing medium to initiate cholesterol depletion.
7. Maintain cells in this condition for up to 7 days, refreshing the medium with pre-warmed MCD-supplemented medium every 2 days to maintain efficient cholesterol depletion.
8. After 7 days of MCD treatment, wash cells gently three times with PBS, and replace with standard DMEM medium (10 % FBS). Incubate for 3−5 days to allow cellular recovery and regrowth.
9. Repeat the entire selection steps (7-day MCD treatment) twice to achieve robust and efficient selection.
10. After screening, extract the genome for next-generation sequencing (ZYMO Quick-DNA Midiprep Plus kit) according to the manufacturer's instructions, and perform comparative analysis of gRNAs between the experimental and control groups.

3.4 Next-generation sequencing (NGS) suquencing and analysis

3.4.1 Library PCR for NGS

1. Amplify the target sequences using synthesized primers with integrated adapters and Hieff Canace® Plus PCR Master Mix (Yeasen), a high-fidelity DNA polymerase premix. The reaction mixture for amplification is composed as follows: 2 ×Hieff Canace® Plus PCR Master Mix, 20 μL; 10 μM Primer F 1.5 μL; 10 μM Primer F 1.5 μL; Template (gDNA) 1 μL; ddH$_2$O 16 μL. PCR Cycling Conditions: 98 °C, 2 min, 1 cycle; 98 °C, 10 s, 58–64 °C, 10 s, 72 °C, 10 s, 8 cycles, each cycle increasing by 1 °C; 98 °C, 10 s, 64 °C, 10 s, 72 °C, 10 s, 30 cycles; 72 °C, 5 min. Primer F: TCGTCGGCAGCGTCAGATGTGTATAA GAGACAGATTTCTTGGGTAGTTTGCAGTTT. Primer R: GTC TCGTGGGCTCGGAGATGTGTATAAGAGACAGGACTCGGTG CCACTTTTTCA.
2. For rapid next-generation sequencing (NGS), 2 μL of PCR product is used as input for library construction using the S-1.2x-Tn5 Tagmentation Mix (a Tn5 transposase-based library preparation system). The tagmentation reaction is carried out according to the manufacturer's instructions.
3. Following tagmentation, the reaction is purified and the library is analyzed by agarose gel electrophoresis (2 % agarose, 1 × TAE buffer). The target DNA fragment range is excised from the gel, and DNA is recovered using a gel extraction kit (e.g., QIAquick Gel Extraction Kit, Qiagen) to remove excess enzymes and short fragments.
4. The purified library is then quantified by Qubit dsDNA HS Assay Kit. Sequencing is performed on the Illumina MiSeq platform using the MiSeq Reagent Micro Kit v2, following the manufacturer's protocols.

3.4.2 Data analysis

Model-based analysis of genome-wide CRISPR/Cas9 knockout (MAGeCK) (Li et al., 2014) can be used to analyze the raw data. The MAGeCK computational pipeline employs a negative binomial model to test for significant differences in sgRNA representation across conditions. Subsequently, it applies a robust ranking aggregation approach to identify genes and pathways that are positively or negatively selected during the screen (Fig. 2).

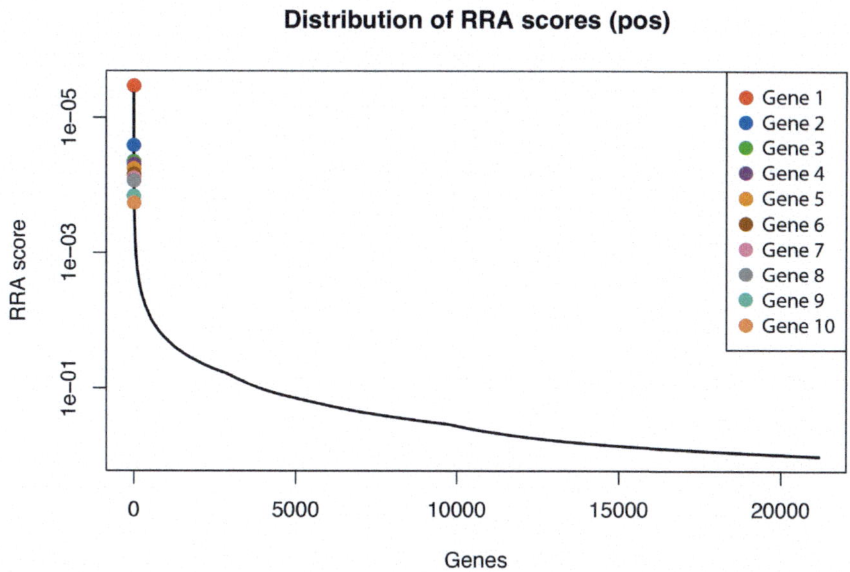

Fig. 2 Representative screening results generated using this protocol. Distribution of RRA scores across the entire genome-wide library. The top ten hit genes identified in the positive selection screen are highlighted. This figure is modified from Joung et al. (2017).

4. Notes

1. A custom sgRNA library can be designed and amplified using the same protocol.
2. Ensuring at least 50 independent clones per sgRNA is critical for preserving complete library representation and preventing sgRNA dropout during amplification processes.
3. Insufficient colony numbers after electroporation are most commonly caused by suboptimal electroporation conditions, low DNA quality or concentration, or salt contamination. All cuvettes, cells, and tubes should be thoroughly chilled, and recovery medium should be added immediately after electroporation to maximize efficiency.
4. Low lentiviral titers may result from inefficient virus production or poor transduction efficiency. In such cases, lentivirus can be concentrated using centrifugal filter units or ultracentrifugation, followed by resuspension in cold PBS and storage at 4 °C overnight to improve recovery.

5. If genome-wide screening is not feasible due to limitations in cell number or viral titer, focused or pathway-specific sgRNA libraries can be used as an alternative approach while maintaining the overall workflow.
6. Antibiotic concentrations should be minimized to avoid excessive selection pressure that may bias enrichment of cells transduced with multiple sgRNAs.
7. A target coverage of > 500 cells per sgRNA is recommended to ensure adequate representation of each perturbation in the final screening readout.
8. Coverage should be increased under suboptimal selection conditions, defined as situations where selective treatments induce only partial cell death, where selection duration is limited, as well as during negative selection screens, in order to preserve sgRNA representation and enhance detection sensitivity.
9. Transducing the sgRNA library at an MOI 0.3 ensures that most cells receive at most one sgRNA, while higher MOIs risk confounding results.
10. For puromycin selection, we recommend that researchers determine the optimal concentration by performing a puromycin kill-curve assay in advance and selecting the minimal dose that results in complete death of uninfected cells, as sensitivities vary between cell types.
11. When utilizing the ZYMO Quick-DNA Midiprep Plus kit, make sure to tighten the reservoir-column connections and centrifuge at optimal speed/duration to eliminate residual buffers. Performing a final dry spin to remove trace wash buffers.
12. To ensure sufficient positive selection efficiency, preliminary experiments are recommended. The duration of drug treatment in each selection round should exceed the time required for complete elimination of wild-type (WT) cells. Prolonged selection (e.g., up to one week) ensures complete elimination of negative cells while allowing sufficient expansion of positive cells, thereby significantly increasing the proportion of positive clones.
13. To minimize amplification errors during sgRNA PCR, a high-fidelity DNA polymerase should be used. Alternative high-fidelity enzymes, including such as NEBNext Ultra II (NEB), PfuUltra II (Agilent) or Kapa HiFi (Roche), are also suitable.

5. Concluding remarks

The protocol described here provides a straightforward approach to identify new candidates involved in cell death pathway. This strategy has recently been used to identify new necroptosis regulators (Callow et al., 2018). This study conducted a CRISPR/Cas9-mediated genome-wide knockout screen to identify regulators of cytokine-induced necroptosis resistance. They demonstrated that the RNA-binding protein PTBP1 was functionally characterized as a critical regulator of necroptosis, maintaining RIPK1 protein abundance through its role in alternative splice-site selection of RIPK1 transcripts. In addition to positive selection screens, negative selection screens have been used to identify pro-survival genes (Huyghe et al., 2022). They performed a pooled genome-wide CRISPR-Cas9-based negative selection screen in MEFs to identify genes that, upon deletion, switch the TNF response from survival to death. They found that autophagy-related 9 A (ATG9A) and 200kD FAK family kinase-interacting protein (FIP200) prevents TNF cytotoxicity via the detoxification of the cytotoxic complex IIa.

Ferroptosis is a form of regulated cell death driven by lipid peroxidation and is modulated by multiple metabolic and genetic pathways (Dixon & Olzmann, 2024). In addition to glutathione-dependent antioxidant systems, recent studies have identified ferroptosis suppressor protein 1 (FSP1/AIFM2) as a glutathione-independent ferroptosis suppressor, highlighting previously unappreciated regulatory mechanisms uncovered through unbiased genetic screening approaches (Doll et al., 2019). Genome-wide CRISPR-Cas9 screening has revealed that enzymes in the distal cholesterol biosynthesis pathway regulate ferroptosis by modulating cellular levels of 7-dehydrocholesterol (7-DHC), a metabolic intermediate produced by SC5D and consumed by DHCR7 during cholesterol synthesis (Li et al., 2024).

Genome-wide CRISPR/Cas9 screening also contributes to dissect the molarcular mechanism in pyroptosis pathway. For example, CRISPR screening has identified gasdermin D as the substrate of inflammatory caspase, which serves as a key execution step in pyroptotic cell death (Shi et al., 2015). Pathogen-derived proteases such as SpeB can trigger keratinocyte pyroptosis in a GSDMA-dependent manner. This mechanism was elucidated using genome-wide CRISPR screens, which identified GSDMA as an essential executioner of SpeB-induced pyroptosis (Deng et al., 2022).

CRISPR screening offers unparalleled comprehensive coverage of the genome. By utilizing sgRNA libraries targeting tens of thousands of genes (e.g., GeCKO v2 with 123,411 sgRNAs), it achieves near-saturation gene knockout in pooled or arrayed formats. This allows unbiased discovery of both canonical and non-canonical pathway components. The ability to generate stable loss-of-function mutations provides a major advantage over RNAi-based approaches, which often suffer from incomplete knockdown and off-target effects. In addition, pooled CRISPR screens allow millions of cells to be profiled in parallel, and are compatible with diverse selection strategies, including drug-induced cytotoxicity and stress-based conditions.

Despite these advantages, several limitations should be considered. Genome-wide CRISPR screening is inherently time-consuming. The workflow typically involves multiple time-consuming steps, including library amplification, large-scale lentivirus production, transduction at sufficient coverage, extended selection periods, next-generation sequencing, and computational analysis. As a result, the overall cost of reagents, sequencing, and infrastructure can be substantial, particularly for genome-wide libraries or replicate screens. In addition, maintaining adequate library coverage requires large numbers of cells, which may limit the applicability of this approach in primary cells, slow-growing cell types, or complex organoid systems, and also increases the overall cost due to large-scale cell culture and maintenance.

In summary, genome-wide CRISPR-Cas9 screening represents a robust and flexible framework for identifying regulators of cell death. Careful consideration of experimental design, selection strategy, and system-specific constraints is essential to maximize the strengths of this approach. This powerful tool would accelerate the discovery of context-specific proteins involved in cell death in future. Future advancements in reducing off-target effects, optimizing in vivo delivery, and refining computational frameworks will further expand its utility in addressing complex biological questions.

Acknowledgments

The work was supported in part by grants from the National Natural Science Foundation of China (32494763), the National Science Fund for Distinguished Young Scholars (32425021), the STI2030-Major Projects (2022ZD0213200), the Talent Plan of Shanghai Branch, Chinese Academy of Sciences (CASSHB-QNPD-2023–010), the Shanghai Science and Technology Development Funds (22JC1410400), the Shanghai Key Laboratory of Aging Studies (19DZ2260400), Shanghai Basic Research Pioneer Project, Shanghai Municipal Science and Technology Major Project, the Postdoctoral Fellowship Program (Grade C) of China Postdoctoral Science Foundation (GZC20232864), and the China Postdoctoral Science Foundation (2023M743667). D.X. is a SANS Exploration Scholar.

Conflicts of interests

The authors have no conflicts of interest to declare.

References

Barrangou, R., & Doudna, J. A. (2016). Applications of CRISPR technologies in research and beyond. *Nature Biotechnology, 34*(9), 933–941. https://doi.org/10.1038/nbt.3659.

Callow, M. G., Watanabe, C., Wickliffe, K. E., Bainer, R., Kummerfield, S., Weng, J., ... Costa, M. R. (2018). CRISPR whole-genome screening identifies new necroptosis regulators and RIPK1 alternative splicing. *Cell Death Disease, 9*, 261. https://doi.org/10.1038/s41419-018-0301-y.

Deng, W., Bai, Y., Deng, F., Pan, Y., Mei, S., Zheng, Z., ... Liu, X. (2022). Streptococcal pyrogenic exotoxin B cleaves GSDMA and triggers pyroptosis. *Nature, 602*(7897), 496–502. https://doi.org/10.1038/s41586-021-04384-4.

Dixon, S. J., & Olzmann, J. A. (2024). The cell biology of ferroptosis. *Nature Reviews. Molecular Cell Biology, 25*(6), 424–442. https://doi.org/10.1038/s41580-024-00703-5.

Doll, S., Freitas, F. P., Shah, R., Aldrovandi, M., da Silva, M. C., Ingold, I., ... Conrad, M. (2019). FSP1 is a glutathione-independent ferroptosis suppressor. *Nature, 575*(7784), 693–698. https://doi.org/10.1038/s41586-019-1707-0.

Garneau, J. E., Dupuis, M.È., Villion, M., Romero, D. A., Barrangou, R., Boyaval, P., ... Moineau, S. (2010). The CRISPR/Cas bacterial immune system cleaves bacteriophage and plasmid DNA. *Nature, 468*(7320), 67–71. https://doi.org/10.1038/nature09523.

Gasiunas, G., Barrangou, R., Horvath, P., & Siksnys, V. (2012). Cas9-crRNA ribonucleoprotein complex mediates specific DNA cleavage for adaptive immunity in bacteria. *Proceedings of the National Academy of Sciences of the United States of America, 109*(39), E2579–E2586. https://doi.org/10.1073/pnas.1208507109.

Huyghe, J., Priem, D., Van Hove, L., Gilbert, B., Fritsch, J., Uchiyama, Y., ... Bertrand, M. J. M. (2022). ATG9A prevents TNF cytotoxicity by an unconventional lysosomal targeting pathway. *Science, 378*, 1201–1207. https://doi.org/10.1126/science.add6967.

Jinek, M., Chylinski, K., Fonfara, I., Hauer, M., Doudna, J. A., & Charpentier, E. (2012). A programmable dual-RNA-guided DNA endonuclease in adaptive bacterial immunity. *Science, 337*(6096), 816–821. https://doi.org/10.1126/science.1225829.

Joung, J., Konermann, S., Gootenberg, J,S., Abudayyeh, O. O., Platt, R. J., Brigham, M. D., ... Zhang, F. (2017). Genome-scale CRISPR-Cas9 knockout and transcriptional activation screening. *Nature Protocols, 12*(4), 828–863. https://doi.org/10.1038/nprot.2017.016.

Li, W., Xu, H., Xiao, T., Cong, L., Love, M. I., Zhang, F., ... Liu, X. S. (2014). MAGeCK enables robust identification of essential genes from genome-scale CRISPR/Cas9 knockout screens. *Genome Biology, 15*(12), 554. https://doi.org/10.1186/s13059-014-0554-4.

Li, Y., Ran, Q., Duan, Q., Jin, J., Wang, Y., Yu, L., ... Wang, P. (2024). 7-Dehydrocholesterol dictates ferroptosis sensitivity. *Nature, 626*(7998), 411–418. https://doi.org/10.1038/s41586-023-06983-9.

Marraffini, L. A., & Sontheimer, E. J. (2008). CRISPR interference limits horizontal gene transfer in staphylococci by targeting DNA. *Science, 322*(5909), 1843–1845. https://doi.org/10.1126/science.1165771.

Shalem, O., Sanjana, N. E., Hartenian, E., Shi, X., Scott, D. A., Mikkelson, T., ... Zhang, F. (2014). Genome-scale CRISPR-Cas9 knockout screening in human cells. *Science, 343*(6166), 84–87. https://doi.org/10.1126/science.1247005.

Shi, J., Zhao, Y., Wang, K., Shi, X., Wang, Y., Huang, H., ... Shao, F. (2015). Cleavage of GSDMD by inflammatory caspases determines pyroptotic cell death. *Nature, 526*(7575), 660–665. https://doi.org/10.1038/nature15514.

Wang, H., La Russa, M., & Qi, L. S. (2016). CRISPR/Cas9 in genome editing and beyond. *Annual Review of Biochemistry, 85*, 227–264. https://doi.org/10.1146/annurev-biochem-060815-014607.

CHAPTER SIX

Flow cytometric distinction of early apoptotic, late apoptotic and primary necrotic morphological markers in *Saccharomyces cerevisiae*

Martin N. Odabas[a], Andreas Zimmermann[a,b,c], Katharina Kainz[a], Maria A. Bauer[a], Tobias Eisenberg[a,b,c], Frank Madeo[a,b,c,*], and Didac Carmona-Gutierrez[a,c,*]

[a]Institute of Molecular Biosciences, NAWI Graz, University of Graz, Graz, Austria
[b]BioTechMed-Graz, Graz, Austria
[c]Field of Excellence BioHealth, University of Graz, Graz, Austria
*Corresponding authors. e-mail address: frank.madeo@uni-graz.at; carmonag@uni-graz.at

Contents

Abstract

Regulated cell death (RCD) is a fundamental biological process that ensures tissue homeostasis, mediates stress responses, and contributes to diverse pathological conditions. The budding yeast *Saccharomyces cerevisiae* has emerged as a valuable model organism for the study of RCD, providing evolutionary insights into conserved molecular pathways and enabling the systematic dissection of cell death mechanisms in a genetically tractable system. Upon a lethal stimulus, yeast populations may segregate into three major subtypes: primary necrotic cells with disrupted plasma membranes, early apoptotic cells exhibiting morphological markers of apoptosis, and late apoptotic (or secondary necrotic) cells that display both apoptotic and necrotic features. Discriminating between these subpopulations is essential for accurate

Methods in Cell Biology, Volume 210
ISSN 0091-679X, https://doi.org/10.1016/bs.mcb.2026.05.018

interpretation of RCD dynamics and has been facilitated by the transition from fluorescence microscopy to flow cytometry, which allows rapid, quantitative, and high-throughput analysis. This manuscript provides a methodological framework for the flow cytometric identification and quantification of these distinct subpopulations of dead or dying yeast, enabling reproducible and detailed assessment of cell death heterogeneity.

1. Introduction

Regulated cell death (RCD) is a genetically controlled and highly orchestrated process, by which cells actively initiate their own demise in response to specific signals. Unlike accidental or uncontrolled cell death, RCD involves defined molecular pathways and specific effectors that are essential for maintaining tissue homeostasis and organismal health (Tang et al., 2019). It plays a central role in various physiological contexts, including embryonic development (e.g., shaping organs and extremities through the elimination of excess cells), immune regulation (e.g., deletion of autoreactive lymphocytes), and protection against infections and malignancies by removing damaged or infected cells (Ellis et al., 1991; Galluzzi, Yamazaki, et al., 2018; Orrenius et al., 2011). Given its integral role in these processes, disruption of RCD pathways can lead to a wide range of pathologies. For instance, excessive RCD contributes to degenerative disorders such as Alzheimer's and Parkinson's disease, while insufficient RCD is a hallmark of cancer and cellular senescence, allowing abnormal cells to evade self-elimination (Tang et al., 2019; Trauth et al., 1989; Vitale et al., 2023). Multiple subroutines of RCD have been described, each with distinct molecular and mechanistic characteristics, e.g., apoptosis, necroptosis, pyroptosis, ferroptosis, or parthanatos (Galluzzi, Vitale, et al., 2018; Yuan & Ofengeim, 2024). Among them, apoptosis was the first to be discovered and remains the most extensively studied, characterized by functional features like the activation of executioner caspases and accompanied by morphological and biochemical correlates such as externalization of phosphatidylserine (PS), cell shrinkage, chromatin condensation, and membrane blebbing without eliciting inflammation (Galluzzi, Vitale, et al., 2018).

The baker's yeast *Saccharomyces cerevisiae* is a fundamental model organism in eukaryotic biology, having provided critical insights into a wide array of cellular processes conserved from yeast to humans. For example, it has been instrumental in elucidating core aspects of cell cycle regulation, autophagy, or intracellular trafficking (Hohmann, 2016). Importantly, *S. cerevisiae* has also emerged as a valuable model for studying RCD, especially apoptosis (Carmona-Gutierrez, Eisenberg, et al., 2010; Madeo et al., 1997). In fact, in some cases, first mechanistic foothold was initially provided by yeast studies, which were later paralleled by findings in mammalian systems, as exemplified by the role in RCD of the ubiquitin segregase Cdc48/p97, involved in proteostasis control (Madeo et al., 1997; Shirogane et al., 1999). Despite being unicellular, yeast exhibits hallmark morphological features of apoptotic cell death observed in higher eukaryotes, such as PS externalization, DNA fragmentation, and chromatin condensation (Carmona-Gutierrez et al., 2018). Furthermore, yeast cells harbor a caspase functional homolog (the metacaspase Yca1) (Madeo et al., 2002), which is activated under specific conditions to execute cell death in a manner reminiscent of apoptosis in higher eukaryotes. While metacaspases have a different cleavage specificity than caspases (they hydrolyze substrates after arginine or lysine, not after aspartate) and are usually active as calcium-dependent monomers (instead of calcium-independent dimers) (Minina et al., 2020), they also share similarities, for instance, regarding their catalytic diad or secondary structures. More importantly, metacaspases share common or at least overlapping substrates with caspases (Brunette et al., 2023; Silva et al., 2011; Sundström et al., 2009), suggesting functional homology. In fact, caspases and metacaspases are mutually exclusive, implying that both may represent evolutionary variants of a shared ancestral function (Carmona-Gutierrez, Fröhlich, et al., 2010).

Besides Yca1, yeast cells possess further conserved molecular components of RCD (Carmona-Gutierrez, Eisenberg, et al., 2010), including mitochondrial effectors like cytochrome *c* (Ludovico et al., 2002), apoptosis-inducing factor 1 (Aif1) (Wissing et al., 2004), the NADH dehydrogenase Nde1 (Saladi et al., 2020), or Endonuclease G (Nuc1) (Büttner et al., 2007). The high conservation of mitochondrial pathways and core cell death motifs suggests that these systems emerged early in the evolution of eukaryotes and were fundamentally maintained. This may involve aspects ranging from the integration and stabilization of endosymbiotic partners to stress management and survival regulation within cellular communities. Importantly, a number of physiological scenarios, in which lethal control plays a significant role, demonstrate that RCD is not merely a correlative phenomenon but can serve adaptive, community-level functions. For instance, RCD plays causal roles in processes such as chronological and replicative aging (Carmona-Gutierrez & Büttner, 2014; Fabrizio et al., 2004; Herker et al., 2004; Laun et al., 2001), failed mating (Severin & Hyman, 2002) or meiosis

(Ahn et al., 2005), and during colony differentiation (Váchová & Palková, 2005), where selective elimination of damaged, aged, or non-functional cells contributes to the overall fitness of the population. Thus, RCD enhances the survival and competitive advantage of the clonal population, underscoring the general view that *S. cerevisiae* – although classically viewed as a unicellular organism – functions as a cooperative cellular community in dense cultures (Büttner et al., 2006; Carmona-Gutierrez & Büttner, 2014).

In *S. cerevisiae*, as in higher eukaryotes (Costigan et al., 2023; Vermes et al., 1995), a pro-apoptogenic stimulus may give rise to three distinct subpopulations (apart from a fourth one, which is comprised of surviving cells): (i) an early apoptotic population showing morphological markers of apoptosis in the absence of membrane rupture, (ii) a late apoptotic or secondary necrotic population displaying both apoptotic and necrotic morphological hallmarks, and (iii) a primary necrotic population characterized by early plasma membrane disruption (Carmona-Gutierrez et al., 2018; Carmona-Gutierrez, Eisenberg, et al., 2010). Thus, besides (i), necrotic subpopulations may appear. The "purely" necrotic population (iii) may arise, for instance, due to cell-to-cell heterogeneity within the culture. This may result in distinct responses to the same stimulus dosage, with some cells engaging apoptosis while others succumbing to (regulated or unregulated) primary necrosis. In fact, this population (or a fraction of it) may represent a regulated necrotic subpopulation, as observed for example during chronological aging (Carmona-Gutiérrez et al., 2011; Eisenberg et al., 2009). Crosstalk between different RCD subroutines may also shape the outcome (Galluzzi, Vitale, et al., 2018). This phenomenon is also present in mammalian systems, where under specific conditions apoptotic stimuli may also give rise to primary necrotic or necroptotic subfractions (Dunai et al., 2012; Yoo et al., 2012). Finally, the secondary necrotic population (ii), displaying morphological hallmarks of both necrosis and apoptosis, corresponds to late apoptotic cells that, eventually acquire necrotic features. In a yeast culture, which naturally lacks efferocytosis, this subpopulation will inevitably emerge after a certain period following cell death (Carmona-Gutierrez et al., 2018; Carmona-Gutierrez, Eisenberg, et al., 2010). Similarly, in mammalian and human cell culture systems where phagocytes (such as macrophages) are absent or not functional, apoptotic cells that are not cleared undergo secondary necrosis, manifesting necrotic membrane permeabilization and eventual loss of integrity

(Sachet et al., 2017). However, in cultured mammalian cells, primary necrotic cells often appear false positive for some of the typically applied apoptotic markers (i.e., Annexin V staining, see below) and may therefore not be easily distinguished from late-apoptotic, secondary necrotic cells. In such cases, assessing defined time points is essential to capture cell death kinetics and subpopulation dynamics.

The dissection of morphologically apoptotic and necrotic subpopulations in *S. cerevisiae* was initially achieved by fluorescence microscopy, using vital dyes, most prominently propidium iodide (PI) to detect membrane rupture and fluorescein isothiocyanate (FITC)-labelled Annexin V to reveal PS externalization (Carmona-Gutierrez et al., 2018; Carmona-Gutierrez, Eisenberg, et al., 2010). Microscopy allowed the recognition of distinct cell death phenotypes at the single-cell level but remained limited in throughput and statistical robustness. The advent and adaptation of flow cytometry for yeast markedly advanced this field, enabling rapid and quantitative discrimination between the above-mentioned subpopulations in large cell cohorts, thus allowing more precise kinetic and dose–response analyses.

In this article, we provide a step-by-step workflow for the flow cytometric distinction of cells displaying early apoptotic, late apoptotic/secondary necrotic and primary necrotic morphological markers. This assay has been successfully applied to characterize, for instance, how various molecular factors (Büttner et al., 2007; Fannjiang et al., 2004; Li et al., 2006; Madeo et al., 2002; Wissing et al., 2004), cellular metabolites (Carmona-Gutierrez et al., 2011; Ludovico et al., 2001; Madeo et al., 1999; Váchová & Palková, 2005), or different pharmacological treatments (Carmona-Gutierrez et al., 2019; Eisenberg et al., 2009; Farrugia et al., 2013) modulate RCD in yeast. The herein presented protocol is designed to guide researchers from sample preparation through staining and gating, enabling reproducible and quantitative resolution of yeast cell death heterogeneity. An important update is the inclusion of doublet discrimination, i.e., the identification and exclusion of events, in which two or more cells pass through the laser beam together, artificially appearing as a single particle. Finally, we also added a complementary fluorescence microscopy protocol for qualitative evaluation. Beyond offering a practical workflow, the protocol also highlights key considerations that shall facilitate its application across a wide range of experimental conditions.

2. Materials

- PBS: 25 mM potassium phosphate buffer, pH 7.0, 0.9 % NaCl (w/v).
- Propidium iodide (PI) solution (see Note 1): 100 μg/mL PI in ddH_2O; protect from light and store at −20 °C.
- Sorbitol-Buffer (Sorbitol-B): 35 mM potassium phosphate buffer, pH 6.8; 0.5 mM $MgCl_2$; 1.2 M sorbitol; can be stored at −20 °C.
- Sorbitol-Incubation buffer (Sorbitol-Inc) containing calcium: 10 mM HEPES, pH 7.4; 140 mM NaCl; 5 mM $CaCl_2$; 0.6 M sorbitol; can be stored at −20 °C.
- Lyticase (see Note 2) (Sigma-Aldrich #L2524; stock solution: 1000 U/mL in H_2O); can be stored at −20 °C.
- Glucuronidase/Arylsulfatase (see Note 3) (Sigma-Aldrich #BGALA-RO); store at 4 °C (see Note 3).
- AnnexinV-FITC (Invitrogen #A13199); can be stored at 4 °C (see Note 4).
- Agarose (VWR #35-1020).
- 76 × 26 mm glass microscope slide (VWR #631-1550).
- 24×50 mm cover glass (VWR #631-1574).
- Fluorescence-grade immersion oil (e.g., Zeiss 518 F #444960-0000-00F).

3. Methods

As mentioned, during early apoptosis, PS becomes externalized from the inner to the outer leaflet of the plasma membrane; PS can be bound by Annexin V specifically and with high affinity in a Ca^{2+}-dependent manner (Meers & Mealy, 1993; Vermes et al., 1995). Conjugation to a fluorophore such as FITC enables fluorescent detection of PS exposure; upon excitation, Annexin V–FITC emits green fluorescence. By contrast, PI is a red-fluorescent, membrane-impermeant nucleic acid dye that can only enter cells once the plasma membrane is disrupted, as occurs in late apoptosis or necrosis. The complementary use of these two fluorochromes enables clear discrimination between surviving cells (Annexin V^-/PI^-), cells with early apoptotic (Annexin V^+/PI^-), late apoptotic or secondary necrotic (Annexin V^+/PI^+), and primary necrotic (Annexin V^-/PI^+) morphological markers, based on their distinct fluorescence emission profiles in separate channels. This dual-fluorescence approach provides a rapid, quantitative, and mechanistically grounded assessment of cell death markers. To tackle cell death progression under a given condition, distinct time points can be determined to depict the kinetics and dynamic changes of different subpopulations. In the following, the required materials for the assay are listed, followed by a detailed, stepwise description of the experimental procedures for both flow cytometry and fluorescence microscopy. Where appropriate, additional notes and remarks are included to provide explanatory context, highlight critical steps, and offer extended observations that may assist in optimizing the protocol or interpreting the results.

3.1 Flow cytometry protocol

1. Harvest 1.5–2 × 10^7 cells by centrifugation at 1100 g for 3 min (see Note 5).
2. Decant/aspirate the supernatant and wash the pellet in 500 μL Sorbitol-B .
3. Resuspend the pellet in 330 μL Sorbitol-B (see Note 6).
4. Spheroplasting (see Note 7):
   a. Add a combination of lyticase and glucuronidase/arylsulfatase; the exact volumes may differ depending on the tested scenario, but usually range from 1−5 μl for lyticase and 15−25 μl for glucuronidase/arylsulfatase (see Note 8).
   b. Vortex briefly. CRITICAL: From this point forward, do not vortex again, as this might cause cell rupture.
   c. Incubate for 1–1.5 h at 28 °C with gentle shaking (see Note 9).
5. Centrifuge briefly (quick spin, 10,000 g, 10 s) in a tabletop centrifuge (see Note 10).
6. Wash the pellet in 500 μL Sorbitol-B, centrifuge again and carefully remove the supernatant (see Note 11)
7. Resuspend the pellet in 30 μL Sorbitol-Inc.
8. Add 3 μL PI and 3 μL Annexin V-FITC (see Note 12).
9. Incubate for 20 min in the dark.
10. Dilute cells with 500 μL Sorbitol-Inc (see Notes 13 and 14).
11. Transfer 250 μL per well for flow cytometry.
12. Flow cytometry settings:
    - Measure Annexin V-FITC fluorescence using excitation by the blue laser (488 nm) and an appropriate emission channel typically used for FITC or GFP (510/30 nm or similar).
    - Measure PI fluorescence using a yellow-green laser (561 nm) and red channel detector (e.g., PerCP-Cy5.5 at 670 nm) to minimize spectral overlap. If a yellow-green laser is unavailable, appropriate emission filters can be used in combination with the blue laser

excitation, but signal compensation according to the instrument manufacturer's guidelines and standard flow cytometry procedures should be applied.

- Use single-stain controls (Annexin V-FITC only, PI only, and unstained) to establish appropriate photomultiplier tube (PMT) settings, population gates, and compensation if necessary (see Note 15).
- Gate distinct populations on a dot plot displaying FITC versus PI fluorescence (see Note 16) (Fig. 1 A and D).

13. The fraction of each population relative to the parent gate/population can then be used for comparison of samples and statistical analysis (Fig. 1D).

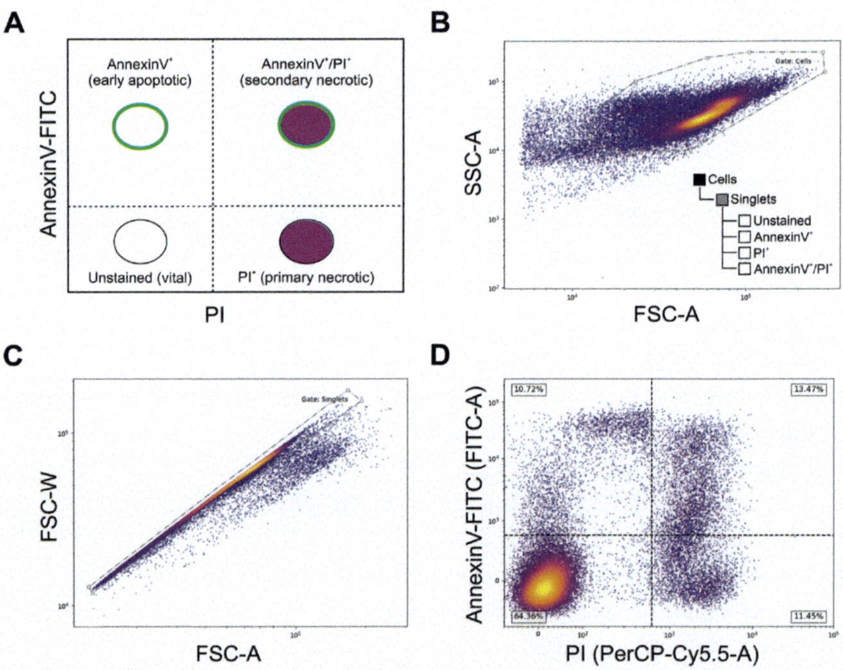

Fig. 1 Flow cytometric analysis of different cell death populations in *S. cerevisiae*. (A) Schematic view of viable (unstained), apoptotic and primary as well as secondary necrotic cells. (B-D) Density plot representation of a 5-day old *S. cerevisiae* BY4742 wildtype batch culture after AnnexinV-FITC/PI co-staining, showing FSC-A vs. SSC-A and the gating hierarchy (B), doublet discrimination (C), and AnnexinV-FITC/PI fluorescence with the population percentages of the parent gate indicated (D). *Plots created with the open-source software EasyFlow. Ma et al. (2024).*

3.2 Fluorescence microscopy protocol

1. Prepare agarose-coated glass slides:
   a. Melt 1 % (w/v) agarose in Sorbitol-Inc using a microwave.
   b. Pipette 1 mL of molten agarose (~40−50 °C) onto a 76 × 26 mm glass microscope slide and allow to solidify at room temperature.
   c. Slides can be stored wrapped in moist paper towels at 4 °C. Equilibrate slides to room temperature for 1 h before use.
2. Perform steps 1–9 as described in the flow cytometry protocol above.
3. Pellet spheroplasts by centrifugation at 500 g for 1 min
4. Transfer 2–3 μL of spheroplasts onto the agarose-coated slide.
5. Carefully place a cover glass on top. CRITICAL: Do not move the cover glass once placed, as this will damage the spheroplasts.
6. Apply a drop of fluorescence-grade immersion oil to the area with the highest spheroplast density.
7. Using an upright widefield fluorescence microscope, focus on an area where cell density is low enough to form a single layer. Typically, 100–200 cells per field of view provide representative results.
8. Acquire transmitted light (DIC) and fluorescence images (AnnexinV-FITC: Ex: 490 nm / Em: 510 nm; PI: Ex: 540–580 nm / Em: 592–668 nm) using a 63 × or 100 × APO 1.4 NA oil immersion objective (see Note 17).

4. Conclusions

S. cerevisiae remains a powerful model organism for exploring the conserved principles of RCD. Its genetic tractability and well-characterized biology make it ideally suited for dissecting the molecular basis of conserved RCD subroutines, including apoptosis. These aspects are relevant not only to higher eukaryotes, including mammals, but also to fungal pathogens, in which RCD mechanisms can be therapeutically exploited to combat infection. The herein described methodology allows for the assessment of distinct morphological features characteristic of apoptotic and necrotic subpopulations (alongside viable cells). Of note, PS externalization remains a morphological indicator of apoptosis and requires functional validation; in fact, the Nomenclature Committee on Cell Death defines apoptosis as a form of cell death driven by caspases, primarily caspase-3 (Galluzzi, Vitale, et al., 2018). We consider metacaspases functional homologs of caspases and interpret Yca1-dependent cell death with apoptotic morphology as functionally apoptotic. When Yca1 is dispensable, the process may still represent an apoptotic subroutine — potentially mediated by other, yet unidentified, caspase-like proteases — or may follow a different regulated pathway. Indeed, although not yet clearly defined, in yeast – as it is the case in mammalian cells (Chen et al., 2019; Li et al., 2015) – PS exposure may not universally accompany apoptosis.

Importantly, the herein presented evaluation of apoptotic and necrotic morphological features provides a starting point to explore the mechanisms underlying a given (genetic, pharmacological or environmental) cell death stimulus. Subsequently, the engagement of specific pathways involved in the identified RCD subroutines may be clarified, e.g., using knockout analyses of key molecular players or monitoring activity of downstream effectors. This also includes distinguishing accidental from regulated necrosis, for instance, by assessing whether the PI-only subpopulation under a given lethal stimulus can be selectively diminished (at least in part) through genetic or pharmacological inhibition, which would indicate the presence of regulatory control. Additional evidence for a regulated form of necrosis could be obtained by examining the nucleo-cytosolic translocation of Nhp6A or the involvement of mitochondrial pathways, both of which have been linked to regulated necrosis in yeast (Carmona-Gutiérrez et al., 2011; Eisenberg et al., 2009; Rockenfeller et al., 2010). An important advantage of these assays is their scalability, making them suitable not only for low-throughput applications but also for high-throughput screening endeavors. As with any modeling approach, however, the findings obtained will need essential validation efforts in the target organism, which may or may not corroborate the initial results observed in *S. cerevisiae*. Nevertheless, the methodology described here has convincingly demonstrated its value as a robust foundation for dissecting apoptotic and necrotic phenotypes.

5. Notes

1. Alternative impermeant DNA-binding dyes can be used, including 7-aminoactinomycin D (7-AAD) or Sytox probes (Riccardi & Nicoletti, 2006; Thakur et al., 2015; Zembruski et al., 2012).
2. Lyticase is an enzyme mixture that degrades the yeast cell wall.
3. Glucuronidase/arylsulfatase catalyzes the simultaneous hydrolysis of β-glucuronides and sulfate esters to facilitate cell wall removal. Do not

store at −20 °C to avoid inactivation of the enzyme mixture. Of note, lyticase has shown superior activity as compared to glucuronidase/arylsulfatase in some settings (e.g., in chronologically aged cells), but empirically testing different combinations of lyticase or glucoronidase/arylsulfatase may be used to optimize conditions for a given cell death trigger.

4. Depending on the manufacturer, different storing conditions may apply (4 °C vs. −20 °C).
5. The typical volume required from a stationary yeast culture at OD_{600} ~3.0 to obtain $1.5–2 \times 10^7$ cells is 150–200 μL.
6. Yeast spheroplasts are osmotically unstable. The addition of sorbitol creates hypertonic conditions, thereby minimizing cell lysis.
7. The yeast cell wall shields the plasma membrane from AnnexinV and thus, it is important to at least partially digest the cell wall to provide accessible PS to AnnexinV before staining.
8. Optimal enzyme concentration and spheroplasting duration depend on experimental conditions and yeast strain. Factors such as nutrient and oxygen availability, pH, temperature, growth kinetics, and genetic mutations can affect cell wall composition and organization. Notably, when cells enter stationary phase, the cell wall becomes thicker, more resistant to β-1,3-glucanase digestion, and less permeable to macromolecules. If the suggested digestion conditions prove insufficient, optimize by increasing enzyme concentrations and/or incubation times. As an example, enzyme application to monitor cells during chronological aging differs between cells younger than 3 days (2.5 μL lyticase and 20 μL glucuronidase/arylsulfatase) and cells older than 3 days (5 μL lyticase and 20 μL glucuronidase/arylsulfatase). To monitor spheroplasting efficiency, measure the decrease in OD_{600} after diluting 10 μL of cells in 1 mL water compared to isotonic buffer. As the cell wall is digested, cells become osmotically sensitive; in water, spheroplasts lyse effectively, resulting in a dramatic decrease in optical density. Conversely, cells with weakened cell walls can be over-digested, so careful optimization is essential. Alternatively, spheroplasting efficiency can also be monitored via microscopy. In this case, the transition from intact, refractile cells to round, phase-dark cells that have lost their rigid cell wall can be observed via bright-field microscopy. Successful spheroplasts typically appear swollen, more uniform, and lack the characteristic bright cell-wall outline seen in untreated cells. At the same time, spheroplasts should not be over-digested and

must maintain cell integrity to ensure the reliability of subsequent analyses. Thus, spheroplasting must be microscopically monitored within short time intervals.

9. Shaking can be performed either in 1.5 mL microcentrifuge tubes using a thermomixer at 200 rpm, or in a Petri dish placed inside a culture shaker at 145 rpm.
10. Cells should not form a tight pellet, since excessive centrifugation may cause spheroplast lysis; instead, a slight smear should be visible on the back side of the reaction tube, a sign of sufficiently moderate centrifugation that also allows spheroplasts to be more easily resuspended.
11. Handle spheroplasts carefully from this point onward to avoid cell damage. Use only 1 mL pipette tips with cut tops for pipetting to prevent shearing. Reaction tubes can be placed up-side down to loosen pellets during washing steps prior to resuspension. Avoid vortexing for spheroplast resuspension.
12. In addition, to probe samples, unstained and single-stained controls should be prepared for flow cytometry settings and population gates.
13. For microscopy, use cells directly after the 20 min incubation without dilution.
14. It is absolutely essential to use a buffer that contains Ca^{2+} ions, which facilitate the interaction of Annexin V with PS.
15. For single-stain controls, select samples with detectable cell death to ensure both positive and negative populations are present. This enables proper adjustment of PMT voltages and accurate gate placement.
16. The population of intact cells should be gated via forward scatter area (FSC-A) and side scatter area (SSC-A) (Fig. 1B). Doublets can lead to false double-positive cells and should be excluded from the analysis using a gate for the linear relationship between FSC-A and either forward scatter height (FSC-H) or alternatively forward scatter width (FSC-W) signals (Fig. 1C).
17. While microscopic analysis is primarily intended for qualitative assessment (flow cytometry provides increased precision, sensitivity, and statistical reliability), it may be employed for quantification when flow cytometric analysis is not feasible. In those cases (a) Overlay the green (Annexin V-FITC) and red (PI) channels, (b) count spheroplasts in a blinded manner, excluding spheroplasts at image edges that are incompletely captured and (c) classify each spheroplast into one of four categories: (i) unstained, (ii) Annexin V-positive only, (iii) PI-positive only, and (iv) Annexin V/PI double-positive (Fig. 2).

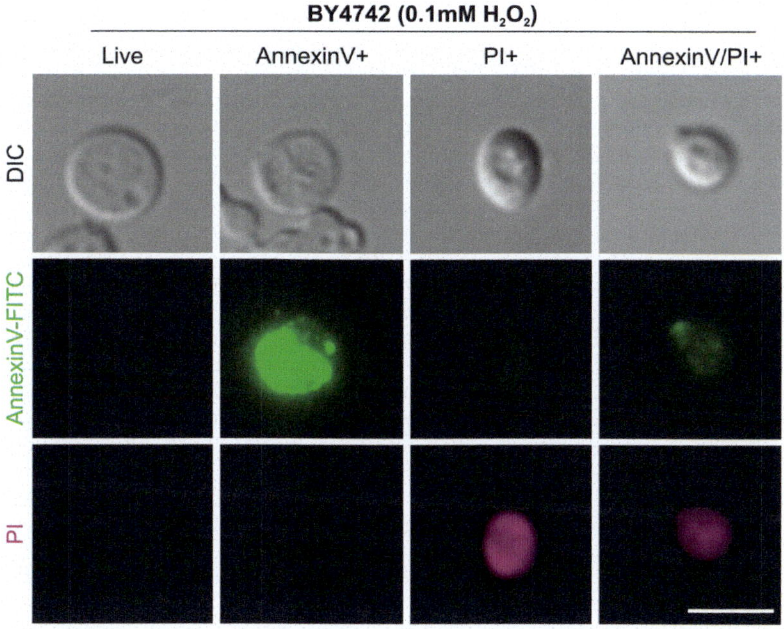

Fig. 2 Yeast cell death phenotyping using fluorescence microscopy. BY4742 wildtype cells grown to early log phase ($OD_{600} = 0.2$) were treated with 0.1 mM H_2O_2 for 4 h at 28 °C and 145 rpm. Cells were stained according to the protocol and then imaged using a Leica DM6B microscope with an HC PL APO 100x/1.40 OIL lens. Images were taken with a Leica-DFC9000GT camera at 16-bit (low noise, high well setting; 6.15 ms for DIC; 50 ms for PI; 500 ms for AnnexinV-FITC). Scale bar = 5 µm.

Acknowledgements

This research was funded in whole, or in part, by the Austrian Science Fund (FWF) [10.55776/COE14], [10.55776/P37278]. We thank the University of Graz (Institute of Molecular Biosciences) for financial support. FM is grateful to the Austrian Science Fund (FWF) for excellence cluster 10.55776/COE14, which was additionally supported by Land Styria, Stadt Graz and University of Graz. DCG, KK and FM thank the FWF for grant 10.55776/P37278 (project no. P37278) as well as for further grants (DOC-50, F3012, W1226, P29203, P29262, P27893, and P31727). For the purpose of open access, the authors have applied a CC BY public copyright license to any author-accepted manuscript version arising from this submission.

Conflict of interest

The authors declare no conflict of interest.

References

Ahn, S.-H., Henderson, K. A., Keeney, S., & Allis, C. D. (2005). H2B (Ser10) phosphorylation is induced during apoptosis and meiosis in S. cerevisiae. *Cell Cycle (Georgetown, Tex.)*, *4*(6), 780–783. https://doi.org/10.4161/cc.4.6.1745.

Brunette, S., Sharma, A., Bell, R., Puente, L., & Megeney, L. A. (2023). Caspase 3 exhibits a yeast metacaspase proteostasis function that protects mitochondria from toxic TDP43 aggregates. *Microbial Cell, 10*(8), 157–169. https://doi.org/10.15698/mic2023.08.801.

Büttner, S., Eisenberg, T., Carmona-Gutierrez, D., Ruli, D., Knauer, H., Ruckenstuhl, C., ... Madeo, F. (2007). Endonuclease G regulates budding yeast life and death. *Molecular Cell, 25*(2), 233–246. https://doi.org/10.1016/j.molcel.2006.12.021.

Büttner, S., Eisenberg, T., Herker, E., Carmona-Gutierrez, D., Kroemer, G., & Madeo, F. (2006). Why yeast cells can undergo apoptosis: Death in times of peace, love, and war. *The Journal of Cell Biology, 175*(4), 521–525. https://doi.org/10.1083/jcb.200608098.

Carmona-Gutierrez, D., & Büttner, S. (2014). The many ways to age for a single yeast cell. *Yeast (Chichester, England), 31*(8), 289–298. https://doi.org/10.1002/yea.3020.

Carmona-Gutiérrez, D., Bauer, M. A., Ring, J., Knauer, H., Eisenberg, T., Büttner, S., ... Madeo, F. (2011). The propeptide of yeast cathepsin D inhibits programmed necrosis. *Cell Death & Disease, 2*(5), e161. https://doi.org/10.1038/cddis.2011.43.

Carmona-Gutierrez, D., Bauer, M. A., Zimmermann, A., Aguilera, A., Austriaco, N., Ayscough, K., ... Madeo, F. (2018). Guidelines and recommendations on yeast cell death nomenclature. *Microbial Cell (Graz, Austria), 5*(1), 4–31. https://doi.org/10.15698/mic2018.01.607.

Carmona-Gutierrez, D., Eisenberg, T., Büttner, S., Meisinger, C., Kroemer, G., & Madeo, F. (2010). Apoptosis in yeast: Triggers, pathways, subroutines. *Cell Death & Differentiation, 17*(5), 763–773. https://doi.org/10.1038/cdd.2009.219.

Carmona-Gutierrez, D., Fröhlich, K.-U., Kroemer, G., & Madeo, F. (2010). Metacaspases are caspases. Doubt no more. *Cell Death and Differentiation, 17*(3), 377–378. https://doi.org/10.1038/cdd.2009.198.

Carmona-Gutierrez, D., Reisenbichler, A., Heimbucher, P., Bauer, M. A., Braun, R. J., Ruckenstuhl, C., ... Madeo, F. (2011). Ceramide triggers metacaspase-independent mitochondrial cell death in yeast. *Cell Cycle (Georgetown, Tex.), 10*(22), 3973–3978. https://doi.org/10.4161/cc.10.22.18212.

Carmona-Gutierrez, D., Zimmermann, A., Kainz, K., Pietrocola, F., Chen, G., Maglioni, S., ... Madeo, F. (2019). The flavonoid 4,4′-dimethoxychalcone promotes autophagy-dependent longevity across species. *Nature Communications, 10*(1), 651. https://doi.org/10.1038/s41467-019-08555-w.

Chen, J., Kuroki, S., Someda, M., & Yonehara, S. (2019). Interferon-γ induces the cell surface exposure of phosphatidylserine by activating the protein MLKL in the absence of caspase-8 activity. *The Journal of Biological Chemistry, 294*(32), 11994–12006. https://doi.org/10.1074/jbc.RA118.007161.

Costigan, A., Hollville, E., & Martin, S. J. (2023). Discriminating between apoptosis, necrosis, necroptosis, and ferroptosis by microscopy and flow cytometry. *Current Protocols, 3*(12), e951. https://doi.org/10.1002/cpz1.951.

Dunai, Z. A., Imre, G., Barna, G., Korcsmaros, T., Petak, I., Bauer, P. I., & Mihalik, R. (2012). Staurosporine induces necroptotic cell death under caspase-compromised conditions in U937 cells. *PLoS One,* 7(7), e41945. https://doi.org/10.1371/journal.pone.0041945.

Eisenberg, T., Knauer, H., Schauer, A., Büttner, S., Ruckenstuhl, C., Carmona-Gutierrez, D., ... Madeo, F. (2009). Induction of autophagy by spermidine promotes longevity. *Nature Cell Biology, 11*(11), 1305–1314. https://doi.org/10.1038/ncb1975.

Ellis, R. E., Yuan, J., & Horvitz, H. R. (1991). Mechanisms and functions of cell death. *Annual Review of Cell Biology,* 7, 663–698. https://doi.org/10.1146/annurev.cb.07.110191.003311.

Fabrizio, P., Battistella, L., Vardavas, R., Gattazzo, C., Liou, L.-L., Diaspro, A., ... Longo, V. D. (2004). Superoxide is a mediator of an altruistic aging program in Saccharomyces cerevisiae. *The Journal of Cell Biology, 166*(7), 1055–1067. https://doi.org/10.1083/jcb.200404002.

Fannjiang, Y., Cheng, W.-C., Lee, S. J., Qi, B., Pevsner, J., McCaffery, J. M., ... Hardwick, J. M. (2004). Mitochondrial fission proteins regulate programmed cell death in yeast. *Genes & Development, 18*(22), 2785–2797. https://doi.org/10.1101/gad.1247904.

Farrugia, G., Bannister, W. H., Vassallo, N., & Balzan, R. (2013). Aspirin-induced apoptosis of yeast cells is associated with mitochondrial superoxide radical accumulation and NAD(P)H oxidation. *FEMS Yeast Research, 13*(8), 755–768. https://doi.org/10.1111/1567-1364.12075.

Galluzzi, L., Vitale, I., Aaronson, S. A., Abrams, J. M., Adam, D., Agostinis, P., ... Kroemer, G. (2018). Molecular mechanisms of cell death: Recommendations of the nomenclature committee on cell death 2018. *Cell Death and Differentiation, 25*(3), 486–541. https://doi.org/10.1038/s41418-017-0012-4.

Galluzzi, L., Yamazaki, T., & Kroemer, G. (2018). Linking cellular stress responses to systemic homeostasis. *Nature Reviews. Molecular Cell Biology, 19*(11), 731–745. https://doi.org/10.1038/s41580-018-0068-0.

Herker, E., Jungwirth, H., Lehmann, K. A., Maldener, C., Fröhlich, K.-U., Wissing, S., ... Madeo, F. (2004). Chronological aging leads to apoptosis in yeast. *The Journal of Cell Biology, 164*(4), 501–507. https://doi.org/10.1083/jcb.200310014.

Hohmann, S. (2016). Nobel yeast research. *FEMS Yeast Research, 16*(8), fow094. https://doi.org/10.1093/femsyr/fow094.

Laun, P., Pichova, A., Madeo, F., Fuchs, J., Ellinger, A., Kohlwein, S., ... Breitenbach, M. (2001). Aged mother cells of Saccharomyces cerevisiae show markers of oxidative stress and apoptosis. *Molecular Microbiology, 39*(5), 1166–1173.

Li, W., Sun, L., Liang, Q., Wang, J., Mo, W., & Zhou, B. (2006). Yeast AMID homologue Ndi1p displays respiration-restricted apoptotic activity and is involved in chronological aging. *Molecular Biology of the Cell, 17*(4), 1802–1811. https://doi.org/10.1091/mbc.E05-04-0333.

Li, Z., Venegas, V., Nagaoka, Y., Morino, E., Raghavan, P., Audhya, A., ... Zhou, Z. (2015). Necrotic cells actively attract phagocytes through the collaborative action of two distinct PS-exposure mechanisms. *PLoS Genetics, 11*(6), e1005285. https://doi.org/10.1371/journal.pgen.1005285.

Ludovico, P., Rodrigues, F., Almeida, A., Silva, M. T., Barrientos, A., & Côrte-Real, M. (2002). Cytochrome c release and mitochondria involvement in programmed cell death induced by acetic acid in Saccharomyces cerevisiae. *Molecular Biology of the Cell, 13*(8), 2598–2606. https://doi.org/10.1091/mbc.e01-12-0161.

Ludovico, P., Sousa, M. J., Silva, M. T., Leão, C. L., & Côrte-Real, M. (2001). Saccharomyces cerevisiae commits to a programmed cell death process in response to acetic acid. *Microbiology (Reading, England), 147*(Pt 9), 2409–2415. https://doi.org/10.1099/00221287-147-9-2409.

Ma, Y., Eizenberg-Magar, I., & Antebi, Y. (2024). EasyFlow: An open-source, user-friendly cytometry analyzer with graphic user interface (GUI). *PLoS One, 19*(11), e0308873. https://doi.org/10.1371/journal.pone.0308873.

Madeo, F., Fröhlich, E., & Fröhlich, K. U. (1997). A yeast mutant showing diagnostic markers of early and late apoptosis. *The Journal of Cell Biology, 139*(3), 729–734. https://doi.org/10.1083/jcb.139.3.729.

Madeo, F., Fröhlich, E., Ligr, M., Grey, M., Sigrist, S. J., Wolf, D. H., & Fröhlich, K. U. (1999). Oxygen stress: A regulator of apoptosis in yeast. *The Journal of Cell Biology, 145*(4), 757–767. https://doi.org/10.1083/jcb.145.4.757.

Madeo, F., Herker, E., Maldener, C., Wissing, S., Lächelt, S., Herlan, M., ... Fröhlich, K. U. (2002). A caspase-related protease regulates apoptosis in yeast. *Molecular Cell, 9*(4), 911–917. https://doi.org/10.1016/s1097-2765(02)00501-4.

Meers, P., & Mealy, T. (1993). Calcium-dependent annexin V binding to phospholipids: Stoichiometry, specificity, and the role of negative charge. *Biochemistry, 32*(43), 11711–11721. https://doi.org/10.1021/bi00094a030.

Minina, E. A., Staal, J., Alvarez, V. E., Berges, J. A., Berman-Frank, I., Beyaert, R., ... Bozhkov, P. V. (2020). Classification and nomenclature of metacaspases and paracaspases: No more confusion with caspases. *Molecular Cell,* 77(5), 927–929. https://doi.org/10.1016/j.molcel.2019.12.020.

Orrenius, S., Nicotera, P., & Zhivotovsky, B. (2011). Cell death mechanisms and their implications in toxicology. *Toxicological Sciences, 119*(1), 3–19. https://doi.org/10.1093/toxsci/kfq268.

Riccardi, C., & Nicoletti, I. (2006). Analysis of apoptosis by propidium iodide staining and flow cytometry. *Nature Protocols, 1*(3), 1458–1461. https://doi.org/10.1038/nprot.2006.238.

Rockenfeller, P., Ring, J., Muschett, V., Beranek, A., Buettner, S., Carmona-Gutierrez, D., ... Madeo, F. (2010). Fatty acids trigger mitochondrion-dependent necrosis. *Cell Cycle (Georgetown, Tex.), 9*(14), 2836–2842. https://doi.org/10.4161/cc.9.14.12267.

Sachet, M., Liang, Y. Y., & Oehler, R. (2017). The immune response to secondary necrotic cells. *Apoptosis: an International Journal on Programmed Cell Death, 22*(10), 1189–1204. https://doi.org/10.1007/s10495-017-1413-z.

Saladi, S., Boos, F., Poglitsch, M., Meyer, H., Sommer, F., Mühlhaus, T., ... Herrmann, J. M. (2020). The NADH dehydrogenase Nde1 executes cell death after integrating signals from metabolism and proteostasis on the mitochondrial surface. *Molecular Cell,* 77(1), 189–202.e6. https://doi.org/10.1016/j.molcel.2019.09.027.

Severin, F. F., & Hyman, A. A. (2002). Pheromone induces programmed cell death in S. cerevisiae. *Current Biology: CB, 12*(7), R233–R235. https://doi.org/10.1016/s0960-9822(02)00776-5.

Shirogane, T., Fukada, T., Muller, J. M. M., Shima, D. T., Hibi, M., & Hirano, T. (1999). Synergistic roles for Pim-1 and c-Myc in STAT3-mediated cell cycle progression and antiapoptosis. *Immunity, 11*(6), 709–719. https://doi.org/10.1016/S1074-7613(00)80145-4.

Silva, A., Almeida, B., Sampaio-Marques, B., Reis, M. I. R., Ohlmeier, S., Rodrigues, F., ... Ludovico, P. (2011). Glyceraldehyde-3-phosphate dehydrogenase (GAPDH) is a specific substrate of yeast metacaspase. *Biochimica et Biophysica Acta, 1813*(12), 2044–2049. https://doi.org/10.1016/j.bbamcr.2011.09.010.

Sundström, J. F., Vaculova, A., Smertenko, A. P., Savenkov, E. I., Golovko, A., Minina, E., ... Bozhkov, P. V. (2009). Tudor staphylococcal nuclease is an evolutionarily conserved component of the programmed cell death degradome. *Nature Cell Biology, 11*(11), 1347–1354. https://doi.org/10.1038/ncb1979.

Tang, D., Kang, R., Berghe, T. V., Vandenabeele, P., & Kroemer, G. (2019). The molecular machinery of regulated cell death. *Cell Research, 29*(5), 347–364. https://doi.org/10.1038/s41422-019-0164-5.

Thakur, S., Cattoni, D. I., & Nöllmann, M. (2015). The fluorescence properties and binding mechanism of SYTOX green, a bright, low photo-damage DNA intercalating agent. *European Biophysics Journal: EBJ, 44*(5), 337–348. https://doi.org/10.1007/s00249-015-1027-8.

Trauth, B. C., Klas, C., Peters, A. M., Matzku, S., Möller, P., Falk, W., ... Krammer, P. H. (1989). Monoclonal antibody-mediated tumor regression by induction of apoptosis. *Science (New York, N. Y.), 245*(4915), 301–305. https://doi.org/10.1126/science.2787530.

Váchová, L., & Palková, Z. (2005). Physiological regulation of yeast cell death in multicellular colonies is triggered by ammonia. *The Journal of Cell Biology, 169*(5), 711–717. https://doi.org/10.1083/jcb.200410064.

Vermes, I., Haanen, C., Steffens-Nakken, H., & Reutelingsperger, C. (1995). A novel assay for apoptosis. Flow cytometric detection of phosphatidylserine expression on early apoptotic cells using fluorescein labelled Annexin V. *Journal of Immunological Methods, 184*(1), 39–51. https://doi.org/10.1016/0022-1759(95)00072-i.

Vitale, I., Pietrocola, F., Guilbaud, E., Aaronson, S. A., Abrams, J. M., Adam, D., ... Galluzzi, L. (2023). Apoptotic cell death in disease-current understanding of the NCCD 2023. *Cell Death and Differentiation, 30*(5), 1097–1154. https://doi.org/10.1038/s41418-023-01153-w.

Wissing, S., Ludovico, P., Herker, E., Büttner, S., Engelhardt, S. M., Decker, T., ... Madeo, F. (2004). An AIF orthologue regulates apoptosis in yeast. *The Journal of Cell Biology, 166*(7), 969–974. https://doi.org/10.1083/jcb.200404138.

Yoo, S. H., Yoon, Y. G., Lee, J. S., Song, Y. S., Oh, J. S., Park, B. S., ... Yoo, Y. H. (2012). Etoposide induces a mixed type of programmed cell death and overcomes the resistance conferred by Bcl-2 in Hep3B hepatoma cells. *International Journal of Oncology, 41*(4), 1443–1454. https://doi.org/10.3892/ijo.2012.1585.

Yuan, J., & Ofengeim, D. (2024). A guide to cell death pathways. *Nature Reviews. Molecular Cell Biology, 25*(5), 379–395. https://doi.org/10.1038/s41580-023-00689-6.

Zembruski, N. C. L., Stache, V., Haefeli, W. E., & Weiss, J. (2012). 7-Aminoactinomycin D for apoptosis staining in flow cytometry. *Analytical Biochemistry, 429*(1), 79–81. https://doi.org/10.1016/j.ab.2012.07.005.

CHAPTER SEVEN

Cytofluorometric analysis of cell death in primary human neutrophils

Fabienne Wagner, Kathrin Stelzner, Aziza Boyny, Rebecca Koch, and Thomas Rudel*
Chair of Microbiology, University of Würzburg, Würzburg, Bavaria, Germany
*Corresponding author. e-mail address: thomas.rudel@uni-wuerzburg.de

Contents

Abstract

Polymorphonuclear neutrophils (PMNs) are among the first effector cells recruited to the sites of infection, where they play a pivotal role in pathogen clearance and the resolution of inflammation. After phagocytosis and elimination of pathogens, PMNs typically undergo apoptosis and are subsequently removed by macrophages via phagocytosis. Due to their short lifespan, the timely clearance of PMNs is essential for maintaining immune homeostasis. However, stimuli, such as cytokines and bacterial

Methods in Cell Biology, Volume 210
ISSN 0091-679X, https://doi.org/10.1016/bs.mcb.2026.05.008

products can prolong PMN survival, and this delayed apoptosis has been implicated in the pathogenesis of chronic inflammation. Elucidating the mechanisms that regulate PMN lifespan is therefore critical for the development of effective therapeutic strategies. Here we present a streamlined protocol for the isolation of primary human PMNs from venous blood and assessing their purity among intact singlet cells by flow cytometry using CD66b surface expression. We also describe methods for evaluating PMN viability with APC Annexin V/7-AAD or SYTOX™ Green staining. The protocol is easily adaptable to other viability dyes, such as propidium iodide, and can be extended to analyze additional PMN surface markers, including CD11b/Integrin alpha M, by flow cytometry

1. Introduction

Polymorphonuclear neutrophils (PMNs) are the first effector cells recruited to sites of inflammation or bacterial invasion, and their role in eliminating pathogens is well established (Kiviat et al., 1990; Mayadas et al., 2014; Patton & Kuo, 1989). They represent the majority of circulating leukocytes and are short-lived (Mayadas et al., 2014). However, their half-life remains a subject of debate. Depending on the system examined (*in vivo*/ *ex vivo*/in circulation / in tissue) half-lives ranging from 8 h to 5 days have been reported (Basu et al., 2002; Bicknell et al., 1994; Dancey et al., 1976; Pillay et al., 2010; Tofts et al., 2011). Upon encountering a pathogen, PMNs initiate phagocytosis (Moore et al., 1978). In the phagosome, microorganisms are killed via the release of antimicrobial proteins, hydrogen peroxide (Babior et al., 1976; Winterbourn et al., 2006) and oxygen radicals generated by the NADPH oxidase (DeLeo et al., 1999) and myeloperoxidase (Chapman et al., 2002; Green et al., 2014). This process, known as oxidative burst, is a hallmark of PMN antimicrobial activity (Holder et al., 2025). After successful resolution of inflammation, PMNs are removed through apoptosis, a form of programmed cell death, and subsequent phagocytosis by macrophages (Brown et al., 2002; Haslett, 1992; Hong et al., 2012; Savill et al., 1989; Shi et al., 2001). Apoptosis of PMNs is a key step in the resolution of inflammation as it prevents the release of pro-inflammatory cytokines (Hengartner, 2000) and promotes the release of anti-inflammatory mediators upon their clearance by macrophages (Savill et al., 2002). Thus, PMN apoptosis and their subsequent phagocytosis are essential for maintaining physiological PMN homeostasis (Dancey et al., 1976; Walker & Willemze, 1980), eliminating pathogens (Simon, 2003), and preventing tissue damage during neutrophil cell death (Savill & Fadok, 2000). During inflammation, PMN apoptosis can be delayed by cytokines, growth factors and epithelial signals

(Akgul et al., 2001; Ginis & Faller, 1997; Watson et al., 1997; Whitlock et al., 2000). This delayed apoptosis contributes to persistent neutrophilic inflammation and has been implicated in several diseases, such as cystic fibrosis, pneumonia and idiopathic fibrosis (Dibbert et al., 1999). Beyond apoptosis, PMNs can undergo alternative forms of cell death (Holder et al., 2025; Zhang et al., 2024). In a process referred to as suicidal NETosis, PMNs release neutrophil extracellular traps (NETs), DNA fibers loaded with antimicrobial proteins, to capture and kill pathogens (Brinkmann et al., 2004; Douda et al., 2015; Fuchs et al., 2007; Papayannopoulos et al., 2010). *In vitro* studies, using isolated human PMNs, and *in vivo* studies of mice infected with *Candida albicans*, revealed the web-like structure of NETs. These NETs contain smooth stretches with a diameter of 15–17 nm, as well as globular domains of 25 nm, which can form aggregates of around 50 nm (Brinkmann et al., 2004; Urban et al., 2009). Excessive NETosis has been linked to pathological conditions, such as sepsis (Clark et al., 2007), vasculitis (Kessenbrock et al., 2009) and thrombosis (Fuchs et al., 2010). PMNs can also undergo pyroptosis, a proinflammatory form of lytic cell death (Dubyak et al., 2023), or necroptosis, a regulated necrosis depending on receptor interacting protein kinase-3 (RIPK3) and mixed lineage kinase domain-like protein (MLKL) (Wang et al., 2018). In contrast to apoptosis, that is a regulated, non-lytic and immunologically form of cell death, necrosis is an unregulated, and thus accidental, lytic and pro-inflammatory form of cell death (Newton et al., 2024). Only recently, necroptosis has been recognized as regulated cell death with necrotic properties (Conrad et al., 2016; Laster et al., 1988). It has been implicated in tissue injury and inflammation and thus infectious or autoimmune diseases, such as cutaneous vasculitis, ulcerative colitis, and psoriasis (Wang et al., 2018). Understanding the regulation of PMN cell death is important for elucidating how dysregulated neutrophil homeostasis contributes to disease. Due to their terminally differentiated state (Mayadas et al., 2014), the study of PMN death requires freshly isolated primary human cells. However, PMN are prone to activation during the isolation process (Dorward et al., 2013; Haslett et al., 1985; Lane et al., 1982; Macey et al., 1995; Youssef et al., 1995). Nevertheless, it is widely accepted that the utilization of a Ficoll gradient for the isolation process is associated with a reduced propensity for PMN activation (Dorward et al., 2013; Hu, 2012). To ensure experimental validity, high-purity neutrophil preparations are essential (Dorward et al., 2013). For quality control, the surface markers CD66b/CEACAM-8, CD16/Fc gamma RIII and CD11b/Integrin alpha M are recommended, as they are consistently expressed on human

PMNs (Lakschevitz et al., 2016). There are other PMN markers, such as CD45 and CD15 amongst others,(Kuijpers et al., 1991; McKenna et al., 2021; Schofield et al., 2024) that are suitable to build a multicolor panel for the determination of PMN purity. Furthermore, PMN purity can be determined based on the absence of other blood cell markers, such as CD14 (Merah-Mourah et al., 2020). Here, we present a streamlined protocol for the isolation of primary human PMNs from fresh venous blood with and subsequent cytofluorometric analysis of their purity based on CD66b/CEACAM-8 expression. We further provide two complementary approaches to quantify PMN cell death: Annexin V/7-AAD staining to specifically assess apoptosis, and SYTOX™ Green staining, to evaluate overall cell death. These protocols are easily adaptable for the cytofluorometric investigation of other PMN cell surface markers as well as for the usage of other cell viability stains.

2. Materials

2.1 Common disposables (*see Note 1*)

- BD Luer-Lok™ syringe, concentric, 10 mL (#613-3931, Avantor)
- Corning® 15 mL Centrifuge Tubes (#430791, Corning)
- Corning® 50 mL Centrifuge Tubes (#430829, Corning)
- Costar® 6-well Clear TC-treated Multiple Well Plates, Individually Wrapped, Sterile (#3516, Corning)
- Costar® 12-well Clear TC-treated Multiple Well Plates, Individually Wrapped, Sterile (#3513, Corning)
- Costar® 24-well Clear TC-treated Multiple Well Plates, Individually Wrapped, Sterile (#3524, Corning)
- Falcon® 5 mL Round Bottom Polystyrene Test Tube, with Cell Strainer Snap Cap (#352235, Corning)
- Neubauer improved Counting chamber, CE-certified (#ZK06; A. Hartenstein Gmbh)
- Pasteur pipettes made of glass, 230 mm length (#PP07, A. Hartenstein Gmbh)
- Reaction tube, 1.5 mL, PP (#72.690.001, Sarstedt AG & Co. KG)
- S-Monovette® Lithium heparin LH, 9 mL, cap orange, (LxØ): 92 × 16 mm, with plastic label (#02.1065, Sarstedt AG & Co. KG)
- Syringe filter, Filtropur S, PES, pore size: 0.2 μm, for sterile filtration (#83.1826.001, Sarstedt AG & Co. KG)

2.2 Cells and reagents (*see Note 1*)

- 5 x PBS: NaCl: 685 mM NaCl, 13.5 mM KCl, 50 mM Na_2HPO_4, 9 mM KH_2PO_4
- BD Pharmingen™ APC Annexin V (#550475, BD Biosciences)
- BD Pharmingen™ 7-Amino-Actinomycin D (7-AAD) (#559925, BD Biosciences)
- CD66b Monoclonal Antibody (G10F5), APC, eBioscience™, Mouse/IgM (#17-0666-42, Thermo Fisher Scientific)
- DPBS, no calcium, no magnesium (#14190169, Gibco)
- Ficoll-Paque™ PLUS density gradient media (#17144003, Cytiva)
- Human Serum, Heat Inactivated, from human male AB plasma, USA origin, sterile-filtered (#H3667, Sigma Aldrich)
- Poly(vinyl alcohol) (#H341584, Sigma Aldrich), used as 1 % solution in 0.85 % sodium chloride
- RPMI 1640 Medium, GlutaMAX™ Supplement (#61870044, Gibco)
- Sterile ultrapure water
- SYTOX™ Green Dead Cell Stain, for flow cytometry (#S34860, Invitrogen)
- Trypan Blue Solution, 0.4 % (#15250061, Gibco)
- Venous blood from healthy donors (*see Note 2)*

2.3 Equipment (*see Note 1*)

- Benchtop centrifuge, such as Megafuge 1.0 (Heraeus Sepatech)
- DM IL inverted contrasting microscope (#090-135006, Leica)
- Flow cytometer equipped with lasers for acquisition of SSC, FSC, BL1(488 nm laser, 530/30 filter), BL3 (488 nm laser, 695/40 filter), RL1 (637 nm laser, 670/14 filter), such as Attune™ NxT Flow Cytometer (Thermo Scientific).
- Humidified cell culture incubator, such as HERAcell™ 240i CO_2 Incubator (#51027450, Thermo Scientific)
- Laboratory biosafety cabinet (Class II), such as HERAsafe Biological Safety Cabinet (#51022712, Thermo Scientific)

2.4 Software (*see Note 1*)

- FlowJo™ v.10.10.0 (FlowJo LLC)
- Excel (Microsoft)
- Prism v. 10.5.0 (GraphPad)

3. Methods

3.1 Isolation of human polymorphonuclear neutrophils (PMNs) from fresh venous blood

1. For each lithium heparin tube with fresh venous blood (~9 mL): add 15 mL Ficoll-Paque™ PLUS to sterile 50 mL conical tube (*see Note 2–5*).
2. Gently layer the blood onto the Ficoll-Paque™ PLUS, ensuring that the phases remain separated (*see Note 6,7*).
3. Centrifuge at 425 × *g* for 30 min without rotor brakes for this step (*see Note 8*).
4. Collect the upper layer containing the autologous plasma pipette (Fig. 1). Use the plasma for coating as described in 3.2 (*see Note 9*).
5. In parallel to the plasma handling, continue with the PMN isolation.
6. To purify the PMNs, aspirate the layers containing the peripheral blood mononuclear cells and the Ficoll-Paque™ PLUS using a sterile Pasteur pipette (*see* Fig. 1 *for the different layers*).
7. Carefully mix the layer containing erythrocytes, granulocytes and red blood cells with 30 mL 1 % polyvinyl alcohol solution by pipetting.

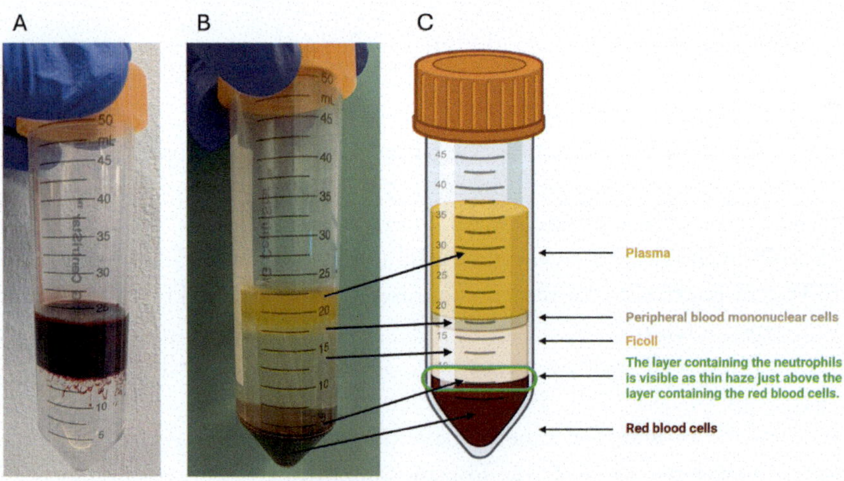

Fig. 1 Blood layers after density centrifugation. After centrifugation at 425 × *g* for 30 min with deactivated rotor brake, the blood separates into different layers. On top is a yellow layer containing plasma followed by a white layer containing the blood mononuclear cells. Below this layer you will find the transparent Ficoll layer. The neutrophils reside in a very thin grey layer just above the layer containing the red blood cells. (A) Before centrifugation. (B) After centrifugation. (C) Schematic overview of the different layers. *Created in BioRender. Wagner (n.d.) https://BioRender.com/wemyw6z.*

8. Incubate the mixture for 45 min on the countertop to completely sediment the red blood cells.
9. Transfer the upper, red layer to a fresh 50 mL conical tube (*see Note 10*).
10. Centrifuge the layer at 188 × *g* for 5 min and remove the supernatant afterwards.
11. Lyse the remaining red blood cells by hypo-osmotic shock: add 16 mL of sterile water to the cells and carefully mix for 40 s (*see Note 11*).
12. Add 4 mL of sterile 5 x PBS to restore osmolarity.
13. Centrifuge at 188 × *g* for 5 min with activated rotor brake and aspirate the supernatant afterwards (*see Note 12*).
14. Re-suspend the PMNs in 5 mL RPMI 1640 Medium, GlutaMAX™ Supplement.
15. Combine cells of one donor if more than one lithium heparin tube with fresh venous blood was used for PMN isolation (*see Note 13).*
16. Determine the PMN density by mixing 10 μL of cells with 10 μL of 0.4 % Trypan Blue solution and counting the cells using a Neubauer improved counting chamber (*see Note 14).*
17. Dilute the PMNs at a density of 1×10^6 cells in RPMI 1640 Medium, GlutaMAX™ Supplement (*see Note 15*).
18. Seed the PMNs at the required amount in 1.5 mL reaction tubes or well plates (*see Note 16–18).* Per condition and staining, it is recommended to use 5×10^5 cells. (*see Note 19)*

3.2 Coating plastic consumables with 10 % autologous plasma

1. Centrifuge the plasma at 755 × *g* for 10 min. There should be little to no pellet visible.
2. Sterile filter the supernatant using a syringe filter with 0.2 μM pore size to ensure that the plasma used for coating is cell-free.
3. Dilute the plasma to 10 % using DPBS.
4. Coat all cell culture plates and tubes used for PMN cultivation with 10 % autologous plasma for 30 min at room temperature (*see Note 9*).
5. Wash the coated plasticware three times with DPBS.
6. Remove the PBS and store the coated plasticware sterile at room temperature until usage.

3.3 CD66b: Staining

1. Transfer 5×10^5 cells to a plasma coated 1.5 mL reaction tube per condition (*see Note 9,16,20*).

2. Centrifuge at 400 × g for 5 min and remove the supernatant afterwards.
3. Staining control: add 100 μL DPBS + 1 % heat-inactivated human serum (*see Note 21,22*).
4. Sample: add 100 μL DPBS + 1 % heat-inactivated human serum + 2.5 μL CD66b monoclonal antibody (*see Note 21–24*).
5. Incubate 30 min in the dark at RT (*see Note 25,26*), then add 500 μL DPBS to each sample.
6. Transfer samples to 5 mL Round Bottom Polystyrene Test Tubes with Cell Strainer.
7. Vortex samples directly before acquisition of fluorescent signals by flow cytometry.

3.4 CD66b: Acquisition and gating

1. Flow cytometry is performed on an instrument equipped with a 637 nm laser.
2. For data acquisition the following parameters are used:
   a. SSC (*see Note 27*)
   b. FSC (*see Note 28*)
   c. RL1 (637 nm, 670/14 filter)
3. The optimal laser voltages (gain) are determined on unstained cells (*see Note 22*).
4. The gates are pre-set on the staining control (*see Note 22*).
5. Per sample, acquire > 10,000 events at a flow rate < 500 events/s (*see Note 29, 30*).
6. The gating is performed as followed (Fig. 2).
   a. Identify the cells on an FSC-H vs SSC-H dot blot. (Fig. 2A).
   b. Identify singlets on cells by using an FSC-H vs FSC-A dot blot. Include events that exhibit a direct proportionality between FSC-H and FSC-A (Fig. 2B).
   c. Identify CD66b^{+} cells on the singlets on an FSC-A vs RL1-A dot blot, based on the staining controls (*see Note 21*) (Fig. 2C).
   d. The data can also be represented as histograms (Fig. 2D).

3.5 APC Annexin V/7-AAD: Staining

1. Transfer 5×10^5 cells to a plasma coated 1.5 mL reaction tube per condition (*see Note 9,20*).
2. Centrifuge at 400 × *g* for 5 min and remove the supernatant afterwards.
3. Staining control: add 100 μL DPBS + 1 % heat-inactivated human serum + 0.2 μL 1 M $CaCl_2$ (*see Note 21,22,31*).

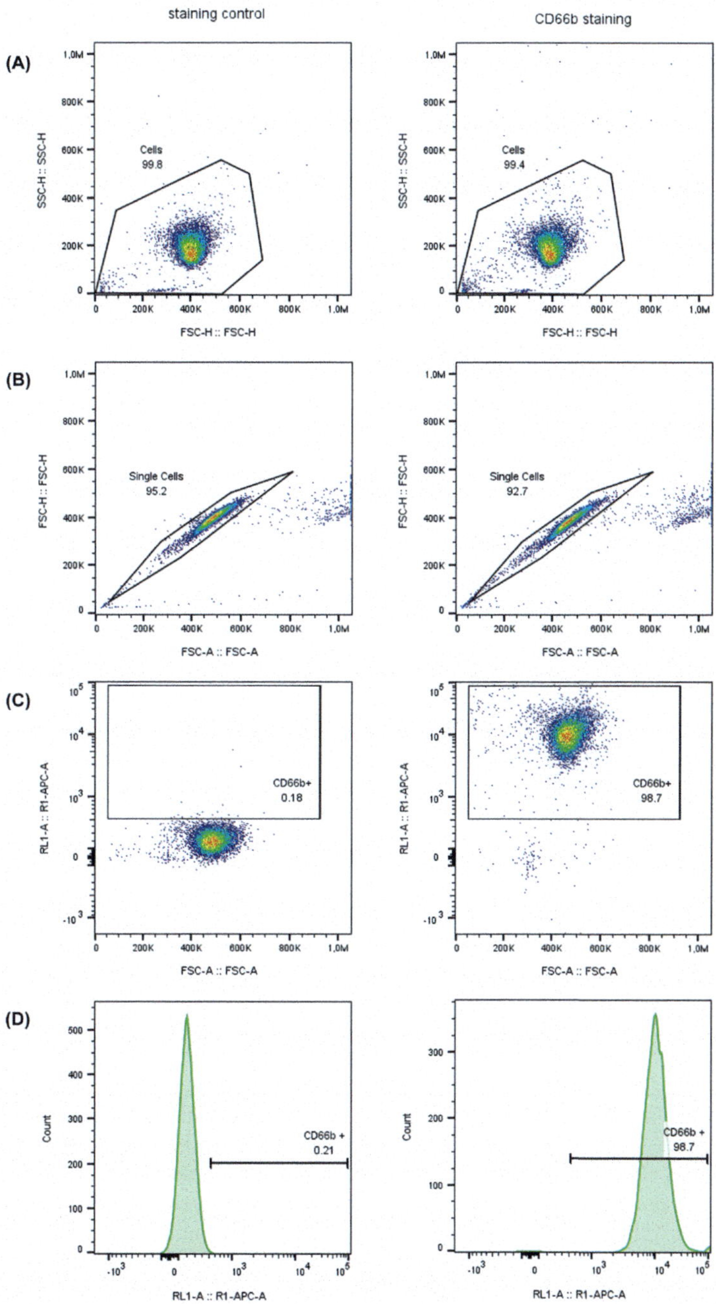

Fig. 2 Analysis of PMN purity using CD66b. PMNs were freshly isolated as described in 3.1 and stained with antibodies specific for CD66b as described in 3.3,

(Continued)

4. For each sample: add 100 µL DPBS + 1 % heat-inactivated human serum + 0.2 µL 1 M $CaCl_2$ + 2.5 µL APC Annexin V + 2.5 µl 7-AAD (*see Note 21–24,31,32*).
5. Incubate 30 min in the dark at RT (*see Note 25,26*), then add 500 µL DPBS to each sample.
6. Transfer samples to 5 mL Round Bottom Polystyrene Test Tubes with Cell Strainer.
7. Vortex samples directly before acquisition of fluorescent signals by flow cytometry.

3.6 APC Annexin V/7-AAD: Acquisition and gating

1. Flow cytometry is performed on an instrument equipped with a 488 nm and a 637 nm laser.
2. For data acquisition the following parameters are used:
   a. SSC (see Note 27)
   b. FSC (see Note 28).
   c. BL3 (488 nm, 695/40 filter)
   d. RL1 (637 nm, 670/14 filter)
3. The optimal laser voltages (gain) are determined on unstained cells (*see Note 22*).
4. The gates are pre-set on the staining controls (*see Note 22*).
5. For each sample, acquire > 10,000 events at a flow rate <500 events/s (*see Note 29, 30*).
6. The gating is performed as followed (Fig. 3).
   e. Identify the cells on an FSC-H vs SSC-H dot blot. (Fig. 3A).
   f. Identify singlets on cells by using an FSC-H vs FSC-A dot blot. Include events that exhibit a direct proportionality between FSC-H and FSC-A (Fig. 3B).
   g. Identify Annexin V + cells on the singlets on an FSC-A vs RL1-A dot blot, based on the staining controls (*see Note 21*) (Fig. 3C).

Fig. 2—Cont'd followed by flow cytometry. The gating strategy was described in 3.4. (A) Identification of cells. (B) Determination of singlets. (C) Identification of $CD66b^+$ cells. (D) Histograms can be used for data representation. The numbers indicate the percentage of events within the gates. The percentages refer to intact singlet neutrophils, gated by FSC/SSC and singlet discrimination, to minimize non-cellular and subcellular artifacts. The representative dot plots and histograms shown here were obtained from one donor to illustrate gating and acquisition. Donor-level replication and statistical summaries are not shown and should be considered when generalizing absolute percentages across populations. Summary statistics across independent donors (n = 5–8) will enable generalization.

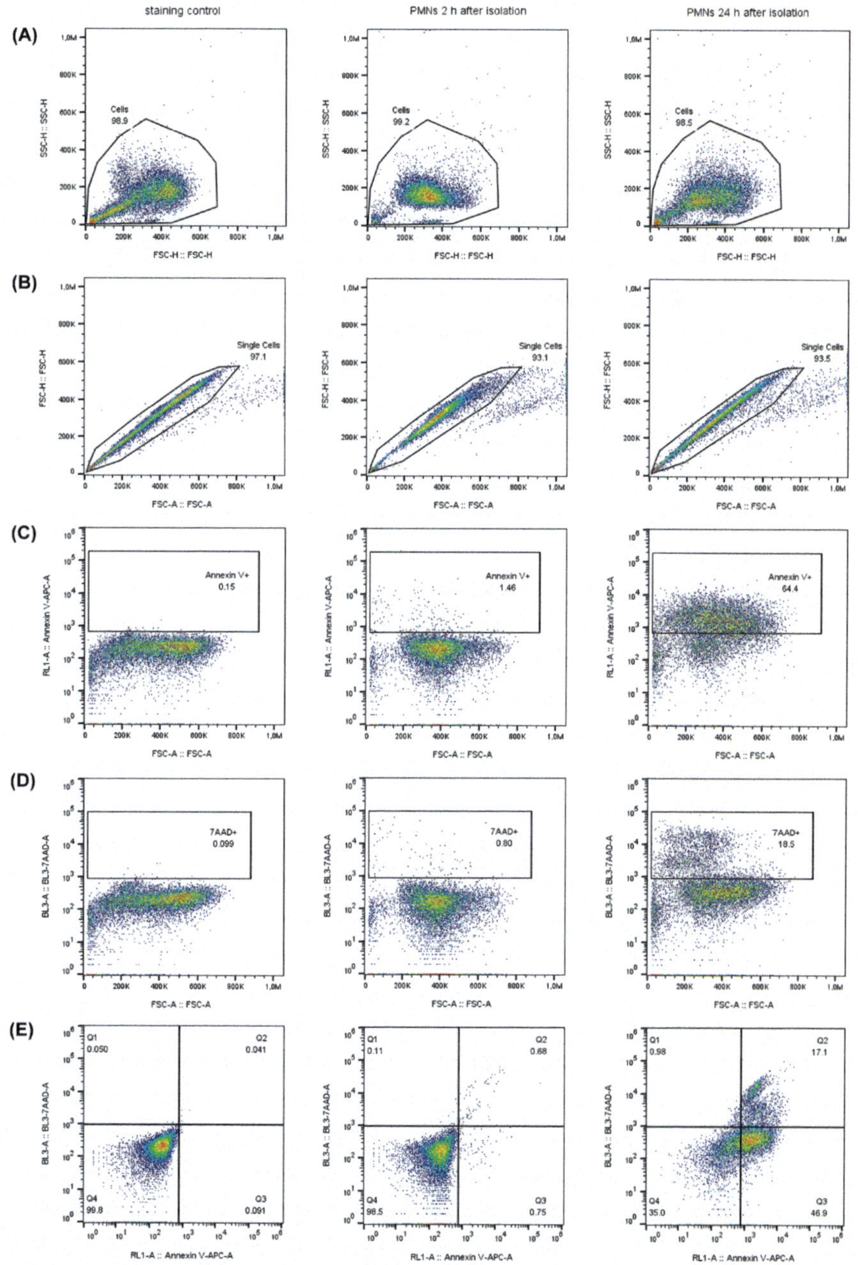

Fig. 3 Analysis of PMN viability using Annexin V and 7-AAD. PMNs were freshly isolated as described in 3.1 and incubated in the presence of Annexin V and 7-AAD as described in 3.5. The gating strategy was described in 3.6. (A) Identification of cells. (B) Determination of singlets. (C) Identification of Annexin V^+ cells. (D) Identification of

(Continued)

h. Identify 7-AAD + cells on the singlets on an FSC-A vs BL3-A dot blot, based on the staining controls (*see Note 22*) (Fig. 3D).

i. Differentiate viable and dead cells on a RL1-A vs BL3-A dot blot (Fig. 3E). Q1: 7-AAD + / Annexin V-: dead cells (compromised cell membrane) Q2: 7-AAD + / Annexin V + : dead cells (late apoptosis, necrosis) Q3: 7-AAD- / Annexin V + : early apoptotic cells Q4: 7-AAD^{-} / Annexin V^{-}: viable cells

3.7 SYTOX™ Green: Staining

1. Transfer 5×10^5 cells to a plasma coated 1.5 mL reaction tube per condition (*see Note 9,20*).
2. Centrifuge at 400 × *g* for 5 min, then remove the supernatant and wash the cells once with DPBS.
3. Centrifuge at 400 × *g* for 5 min and remove the supernatant afterwards.
4. Staining control: add 500 μL DPBS + 1 % heat-inactivated human serum (*see Note 21,22*).
5. For each sample: add 500 μL DPBS + 1 % heat-inactivated human serum + 30 nM SYTOX™ Green (*see Note 21–24*)
6. Incubate 15 min in the dark at RT (*see Note 25,26*).
7. Transfer samples to 5 mL Round Bottom Polystyrene Test Tubes with Cell Strainer.
8. Vortex samples directly before acquisition of fluorescent signals by flow cytometry.

3.8 SYTOX™ Green: Acquisition and gating

1. Flow cytometry is performed on an instrument equipped with a 488 nm laser.
2. For data acquisition the following channels are used:

Fig. 3—Cont'd 7-AAD^{+} cells. (E) Differentiation of viable and dead cells Q1: 7-AAD^{+} / Annexin V^{-}, Q2: 7-AAD^{+} / Annexin V^{+}, Q3: 7-AAD^{-} / Annexin V^{+}, Q4: 7-AAD^{-} / Annexin V^{-}. The numbers indicate the percentage of events within the gates. The percentages refer to intact singlet neutrophils, gated by FSC/SSC and singlet discrimination, to minimize non-cellular and subcellular artifacts. Because the debris and singlet gates exclude low-scatter particles and aggregates, apoptotic bodies and NET-associated material are not enumerated by default and may lead to underestimation of total neutrophil-derived material, particularly at later time points. The representative dot blots and histograms shown here were obtained from one donor to illustrate gating and acquisition. Donor-level replication and statistical summaries are not shown here and should be considered when generalizing absolute percentages across populations or time-of-day. Summary statistics across independent donors (5–8) will enable generalization.

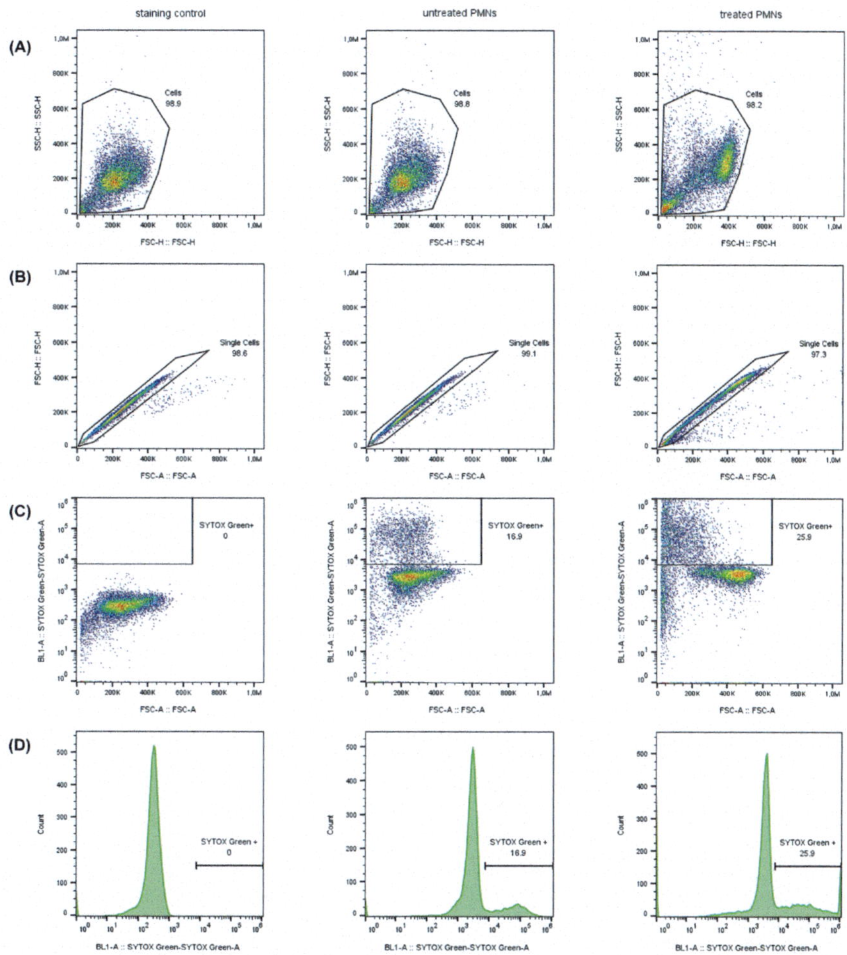

Fig. 4 Analysis of PMN viability using SYTOX™ Green. PMNs were freshly isolated as described in 3.1 and incubated in the presence of SYTOX™ Green as described in 3.7. The gating strategy was described in 3.8. (A) Identification of cells. (B) Determination of singlets. (C) Identification of SYTOX™ Green^{+} cells. (D) Histograms can be used for data representation. The numbers indicate the percentage of events within the gates. The percentages refer to intact singlet neutrophils, gated by FSC/SSC and singlet discrimination, to minimize non-cellular and subcellular artifacts. Because the debris and singlet gates exclude low-scatter particles and aggregates, apoptotic bodies and NET-associated material are not enumerated by default and may lead to underestimation of total neutrophil-derived material, particularly at later time points. The representative dot blots and histograms shown here were obtained from one donor to illustrate gating and acquisition. Donor-level replication and statistical summaries are not shown here and should be considered when generalizing absolute percentages across populations. Summary statistics across independent donors (5–8) will enable generalization.

   a. SSC (*see Note 27*)
   b. FSC (*see Note 28*).
   c. BL1 (488 nm, 530/30 filter)
3. The optimal laser voltages (gain) are determined on unstained cells (*see Note 22*).
4. The gates are pre-set on the staining control (*see Note 22*).
5. For each sample, acquire > 10,000 events at a flow rate <500 events/s (*see Note 29,30*).
6. The gating is performed as followed (Fig. 4)
   a. Identify the cells on an FSC-H vs SSC-H dot blot. (Fig. 4A).
   b. Identify singlets on cells by using an FSC-H vs FSC-A dot blot. Include events that exhibit a direct proportionality between FSC-H and FSC-A (Fig. 4B).
   c. Identify SYTOX™ Green$^+$ cells on the singlets on an FSC-A vs BL1-A dot blot, based on the staining controls (*see Note 21*) (Fig. 4C).
   d. The data can also be represented as histograms (Fig. 4D).

4. Notes

1. Virtually equivalent disposable, reagents, instruments and software can be purchased from multiple other providers at a comparable cost.
2. The venous blood was obtained from healthy donors who signed their consent beforehand. The usage of human blood and neutrophils was approved by the Ethics Commission of the University of Würzburg (Code 300/21).
3. The sample volume is limited to 9 mL blood per S-Monovette® Lithium heparin tube, and it is not recommended to obtain more than 36 mL blood per donor per donation. Sodium heparin tubes can also be used for PMN isolation (Fletcher & Gallin, 1983; Kuang et al., 2023).
4. All steps during the PMN isolation are performed at room temperature.
5. It is recommended to isolate the PMNs as fast as possible after the blood donation, otherwise neutrophil viability might be compromised. It is recommended to perform PMN isolations always at the same time of day to exclude effects of circadian oscillations (Ovadia et al., 2023).
6. If the blood and the Ficoll-Paque™ PLUS are mixed, the experiment is compromised.

7. To avoid mixing of the blood and the Ficoll-Paque™ PLUS, electronic pipet-aids should be placed in drop-wise mode.
8. It is important to deactivate the brake during Ficoll separation to avoid mixing of the phases due to harsh deceleration. For the following centrifugation steps the brake should be reactivated.
9. It is recommended to coat the plasticware used for PMN cultivation with 10 % autologous plasma to minimize activation of the PMNs by the plastic surface.
10. Avoid transferring excess red blood cells and rather leave a bit of supernatant in the tube.
11. It is important not to exceed 40 seconds of incubation time during the water lysis to avoid PMN lysis.
12. If the cell pellet is still red after one round of water lysis, repeat the water lysis. However, it is not recommended to repeat it more than once due to loss in the number of recovered PMNs.
13. We recommend combining cells only if they belong to the same donor and the same blood donation as there is a high variability between the donors (Kippner et al., 2014), especially between donors of different biological sex (Gupta et al., 2020), but also between different donations of the same donor.
14. Cell counting can also be performed using a Burker chamber or a benchtop cell counter such as the Countess 3 Automated Cell Counter.
15. PMNs are cultivated without serum to avoid activation.
16. PMNs can be cultivated in 1.5 mL reaction tubes and 24- or 12-well plates at a maximum amount of 1×10^6 cells. If larger cell numbers are required, it is recommended to culture the PMNs in 6-well plates.
17. It is recommended to rest the PMNs for 30 min at RT on the countertop before starting treatments or other experimental approaches.
18. PMNs are cultured at 37 °C in a humidified atmosphere containing 5 % CO_2.
19. It is recommended to assess PMN purity directly after their isolation. Investigation of cell death can be performed at the desired time points. For example, it has been demonstrated that, following a 24-hour infection with *Chlamydia trachomatis*, approximately 90 % of PMNs remain viable, compared to around 60 % of uninfected PMNs after 24 h. Following a 72-hour cultivation period, approximately 20 % of PMNs remain viable (Koch et al., 2025).

20. PMNs are cultured as suspension and only stick to plastic surfaces when they are activated. However, they can be removed from the surfaces by careful pipetting. It is recommended to confirm the removal from the plate by microscopy.
21. Prior to first use, it is important to inactivate the complement system (Mastellos et al., 2024). Therefore, human serum must be decomplemented at 56 °C for 30 min in a water bath.
22. To adjust the gain, to perform compensation and to set the gates it is crucial to include unstained control cells, as well as samples stained with single fluorophores and *Fluorescence Minus One* (FMO) samples in case several fluorophores are used. FMO samples are generated by staining a sample with all used fluorophores, except for one, to determine spectral overlaps and background signals.
23. These volumes are used per sample, but it is recommended to collectively prepare the staining solution for all required samples plus 10 % extra volume to achieve homogenous staining results.
24. It is recommended to start with fluorophore concentrations as stated by the manufacturer. Titration can be performed to obtain optimal results.
25. The samples should be incubated in the dark to avoid photobleaching of the fluorophore, which could negatively impact the acquisition.
26. It is recommended to incubate the samples for 15–30 min.
27. SSC: side scatter, the detector measures scatter at a ninety-degree angle relative to the laser (information about the granularity)
28. FSC: forward scatter, the detector measures along the laser (information about the size).
29. For gating and analysis, the acquisition of a high number of cells is important. For all samples equal numbers of cells should be measured at a constant flow rate to avoid discrepancies due to technical differences.
30. To obtain the best signal quality a lower flow rate should be used.
31. APC Annexin V binding to phosphatidylserine is calcium dependent.
32. Instead of 7-AAD, propidium iodide can be used as well.

5. Concluding remarks

Delayed PMN apoptosis has been linked to neutrophilic inflammation and thereby is associated with a range of inflammatory and malignant diseases (Dibbert et al., 1999; Zhang et al., 2024). Moreover, prolonged PMN survival has been correlated with persistent bacterial infections (Cherla et al., 2018; Faris et al., 2025; Koch et al., 2025; Rajeeve et al., 2018; Saba et al., 2002). Pathogens exploit several strategies to delay PMN apoptosis, including the activation of pro-survival signaling pathways, inhibition of caspase activity, and stabilization of mitochondrial integrity (Aga et al., 2002; Niu et al., 2010; Wang et al., 2022). For example, the tick-borne bacterium *Anaplasma phagocytophilum* inhibits mitochondria-mediated apoptosis via Ats-1 (Anaplasma translocated substrate 1) thereby preventing cytochrome c release (Niu et al., 2010). Similarly, reduced PMN apoptosis has been linked to decreased caspase-3 activity during *Leishmania major* (Aga et al., 2002) and *Treponema pallidum* (Wang et al., 2022) infections. We have extensively investigated the interaction of the obligate intracellular human pathogen *Chlamydia trachomatis* with primary human PMNs, revealing that chlamydial inhibition of PMN cell death represents a complex, multifactorial process. First, the secreted chlamydial effector CPAF (chlamydial protease-like activating factor) has been shown to suppress PMN effector functions through cleavage of FPR2 (formyl peptide receptor 2), a mechanism associated with extended PMN survival (Rajeeve et al., 2018). In addition, CT181 has been identified as a novel chlamydial effector that binds to and stabilizes anti-apoptotic protein Mcl-1 (induced myeloid leukemia cell differentiation protein), thereby promoting neutrophil longevity (Faris et al., 2025). More recently, activation of the PI3K/Akt survival pathway has been demonstrated to inhibit both apoptosis, via caspase-3 inhibition, and necroptosis, via Mcl-1 stabilization. This Mcl-1 stabilization has been shown to reduce RIP3 (receptor-interacting protein kinase 3) expression, thereby sustaining PMN survival during chlamydial infection (Koch et al., 2025). Here we provide a robust yet straightforward protocol for the isolation of human primary PMNs from peripheral blood. It has been recommended to assess PMN purity via analysis of the expression of multiple surface markers such as CD66b, CD16 and CD11b (Lakschevitz et al., 2016). Therefore, we chose to provide the protocol for the subsequent cytofluorometric assessment of PMN purity through CD66b surface expression as example. The protocol for antibody staining and flow cytometry-assisted analysis can be easily adapted to the analysis of a multicolor panel of other surface markers, such as CD11b/Integrin alpha M or CD35/complement receptor type 1. Furthermore, it is recommended to assess the absence of monocyte markers, such as CD14 and CD3 (Merah-Mourah et al., 2020). We also detail two complementary methods for assessing PMN viability. General cell death can be simply investigated via SYTOX™ Green staining, while apoptosis is monitored by Annexin V/7-AAD staining. The protocols are also compatible with other dyes, such as propidium iodide. Despite their simplicity and ease of implementation, these methodologies have proven instrumental in uncovering key mechanisms of *C. trachomatis*-mediated modulation of PMN survival.

Acknowledgments

This work was supported by the Deutsche Forschungsgemeinschaft (DFG) through the Research Training Group RTG2581, project DFG RU 631/17–1, and by the European Research Council (ERC) under the Advanced Grant ERC-2018-ADG/NCI-CAD to T.R.

Competing interests

The authors declare no conflict of interest.

References

Aga, E., Katschinski, D. R. M., van Zandbergen, G., Laufs, H., Hansen, B., Müller, K., ... Laskay, T. S. (2002). Inhibition of the Spontaneous apoptosis of neutrophil granulocytes by the intracellular parasite leishmania major1. *The Journal of Immunology, 169*(2), 898–905. https://doi.org/10.4049/jimmunol.169.2.898.

Akgul, C., Moulding, D. A., & Edwards, S. W. (2001). Molecular control of neutrophil apoptosis. *FEBS Letters, 487*(3), 318–322. https://doi.org/10.1016/S0014-5793(00)02324-3.

Babior, B. M., Curnutte, J. T., & McMurrich, B. J. (1976). The particulate superoxide-forming system from human neutrophils. Properties of the system and further evidence supporting its participation in the respiratory burst. *The Journal of Clinical Investigation, 58*(4), 989–996. https://doi.org/10.1172/jci108553.

Basu, S., Hodgson, G., Katz, M., & Dunn, A. R. (2002). Evaluation of role of G-CSF in the production, survival, and release of neutrophils from bone marrow into circulation. *Blood, 100*(3), 854–861. https://doi.org/10.1182/blood.V100.3.854.

Bicknell, S., van Eeden, S., Hayashi, S., Hards, J., English, D., & Hogg, J. C. (1994). A non-radioisotopic method for tracing neutrophils in vivo using 5'-bromo-2'-deoxyuridine. *American Journal of Respiratory Cell and Molecular Biology, 10*(1), 16–23. https://doi.org/10.1165/ajrcmb.10.1.8292377.

Brinkmann, V., Reichard, U., Goosmann, C., Fauler, B., Uhlemann, Y., Weiss, D. S., ... Zychlinsky, A. (2004). Neutrophil extracellular traps kill bacteria. *Science (New York, N. Y.), 303*(5663), 1532–1535. https://doi.org/10.1126/science.1092385.

Brown, S., Heinisch, I., Ross, E., Shaw, K., Buckley, C. D., & Savill, J. (2002). Apoptosis disables CD31-mediated cell detachment from phagocytes promoting binding and engulfment. *Nature, 418*(6894), 200–203. https://doi.org/10.1038/nature00811.

Chapman, A. L., Hampton, M. B., Senthilmohan, R., Winterbourn, C. C., & Kettle, A. J. (2002). Chlorination of bacterial and neutrophil proteins during phagocytosis and killing of Staphylococcus aureus. *The Journal of Biological Chemistry, 277*(12), 9757–9762. https://doi.org/10.1074/jbc.M106134200.

Cherla, R., Zhang, Y., Ledbetter, L., & Zhang, G. (2018). Coxiella burnetii inhibits neutrophil apoptosis by exploiting survival pathways and antiapoptotic protein Mcl-1. *Infection and Immunity, 86*(4), https://doi.org/10.1128/iai.00504-17.

Clark, S. R., Ma, A. C., Tavener, S. A., McDonald, B., Goodarzi, Z., Kelly, M. M., ... Kubes, P. (2007). Platelet TLR4 activates neutrophil extracellular traps to ensnare bacteria in septic blood. *Nature Medicine, 13*(4), 463–469. https://doi.org/10.1038/nm1565.

Conrad, M., Angeli, J. P. F., Vandenabeele, P., & Stockwell, B. R. (2016). Regulated necrosis: Disease relevance and therapeutic opportunities. *Nature Reviews. Drug Discovery, 15*(5), 348–366. https://doi.org/10.1038/nrd.2015.6.

Dancey, J. T., Deubelbeiss, K. A., Harker, L. A., & Finch, C. A. (1976). Neutrophil kinetics in man. *The Journal of Clinical Investigation, 58*(3), 705–715. https://doi.org/10.1172/jci108517.

DeLeo, F. R., Allen, L. A., Apicella, M., & Nauseef, W. M. (1999). NADPH oxidase activation and assembly during phagocytosis. *Journal of Immunology, 163*(12), 6732–6740.

Dibbert, B., Weber, M., Nikolaizik, W. H., Vogt, P., Schöni, M. H., Blaser, K., & Simon, H. U. (1999). Cytokine-mediated Bax deficiency and consequent delayed neutrophil apoptosis: A general mechanism to accumulate effector cells in inflammation. *Proceedings of the National Academy of Sciences of the United States of America, 96*(23), 13330–13335. https://doi.org/10.1073/pnas.96.23.13330.

Dorward, D. A., Lucas, C. D., Alessandri, A. L., Marwick, J. A., Rossi, F., Dransfield, I., ... Rossi, A. G. (2013). Technical advance: Autofluorescence-based sorting: Rapid and nonperturbing isolation of ultrapure neutrophils to determine cytokine production. *Journal of Leukocyte Biology, 94*(1), 193–202. https://doi.org/10.1189/jlb.0113040.

Douda, D. N., Khan, M. A., Grasemann, H., & Palaniyar, N. (2015). SK3 channel and mitochondrial ROS mediate NADPH oxidase-independent NETosis induced by calcium influx. *Proceedings of the National Academy of Sciences of the United States of America, 112*(9), 2817–2822. https://doi.org/10.1073/pnas.1414055112.

Dubyak, G. R., Miller, B. A., & Pearlman, E. (2023). Pyroptosis in neutrophils: Multimodal integration of inflammasome and regulated cell death signaling pathways. *Immunological Reviews, 314*(1), 229–249. https://doi.org/10.1111/imr.13186.

Faris, R., Koch, R., McCaslin, P., Challagundla, N., Steiert, B., Andersen, S. E., ... Weber, M. M. (2025). The Chlamydia trachomatis secreted effector protein CT181 binds to Mcl-1 to prolong neutrophil survival. *mBio, 0*, e00357-26. 2025.2003.2016.643443. https://doi.org/10.1101/2025.03.16.643443.

Fletcher, M. P., & Gallin, J. I. (1983). Human neutrophils contain an intracellular pool of putative receptors for the chemoattractant N-formyl-methionyl-leucyl-phenylalanine. *Blood, 62*(4), 792–799. https://doi.org/10.1182/blood.V62.4.792.792.

Fuchs, T. A., Abed, U., Goosmann, C., Hurwitz, R., Schulze, I., Wahn, V., ... Zychlinsky, A. (2007). Novel cell death program leads to neutrophil extracellular traps. *The Journal of Cell Biology, 176*(2), 231–241. https://doi.org/10.1083/jcb.200606027.

Fuchs, T. A., Brill, A., Duerschmied, D., Schatzberg, D., Monestier, M., Myers, D. D., Jr., ... Wagner, D. D. (2010). Extracellular DNA traps promote thrombosis. *Proceedings of the National Academy of Sciences of the United States of America, 107*(36), 15880–15885. https://doi.org/10.1073/pnas.1005743107.

Ginis, I., & Faller, D. V. (1997). Protection from apoptosis in human neutrophils is determined by the surface of adhesion. *American Journal of, 272*(1), C295–C309. https://doi.org/10.1152/ajpcell.1997.272.1.C295.

Green, J. N., Kettle, A. J., & Winterbourn, C. C. (2014). Protein chlorination in neutrophil phagosomes and correlation with bacterial killing. *Free Radical Biology & Medicine, 77*, 49–56. https://doi.org/10.1016/j.freeradbiomed.2014.08.013.

Gupta, S., Nakabo, S., Blanco, L. P., O'Neil, L. J., Wigerblad, G., Goel, R. R., ... Kaplan, M. J. (2020). Sex differences in neutrophil biology modulate response to type I interferons and immunometabolism. *Proceedings of the National Academy of Sciences, 117*(28), 16481–16491. https://doi.org/10.1073/pnas.2003603117.

Haslett, C. (1992). Resolution of acute inflammation and the role of apoptosis in the tissue fate of granulocytes. *Clinical Science, 83*(6), 639–648. https://doi.org/10.1042/cs0830639.

Haslett, C., Guthrie, L. A., Kopaniak, M. M., Johnston, R. B., Jr., & Henson, P. M. (1985). Modulation of multiple neutrophil functions by preparative methods or trace concentrations of bacterial lipopolysaccharide. *The American Journal of Pathology, 119*(1), 101–110.

Hengartner, M. O. (2000). The biochemistry of apoptosis. *Nature, 407*(6805), 770–776. https://doi.org/10.1038/35037710.

Holder, B. E., Reber, C. P., & Monteith, A. J. (2025). Live fast, die young: Neutrophils streamline their metabolism to maximize inflammation. *Infection and Immunity, 93*(8), e00498–00424. https://doi.org/10.1128/iai.00498-24.

Hong, C., Kidani, Y., N, A. G., Phung, T., Ito, A., Rong, X., ... Bensinger, S. J. (2012). Coordinate regulation of neutrophil homeostasis by liver X receptors in mice. *The Journal of Clinical Investigation, 122*(1), 337–347. https://doi.org/10.1172/jci58393.

Hu, Y. (2012). Isolation of human and mouse neutrophils ex vivo and in vitro. In R. B. Ashman (Ed.). *Leucocytes: Methods and protocols* (pp. 101–113). Humana Press. https://doi.org/10.1007/978-1-61779-527-5_7.

Kessenbrock, K., Krumbholz, M., Schönermarck, U., Back, W., Gross, W. L., Werb, Z., ... Jenne, D. E. (2009). Netting neutrophils in autoimmune small-vessel vasculitis. *Nature Medicine, 15*(6), 623–625. https://doi.org/10.1038/nm.1959.

Kippner, L. E., Kim, J., Gibson, G., & Kemp, M. L. (2014). Single cell transcriptional analysis reveals novel innate immune cell types. *PeerJ, 2*, e452. https://doi.org/10.7717/peerj.452.

Kiviat, N. B., Paavonen, J. A., Wølner-Hanssen, P., Critchlow, C. W., Stamm, W. E., Douglas, J., ... Holmes, K. K. (1990). Histopathology of endocervical infection caused by Chlamydia trachomatis, herpes simplex virus, Trichomonas vaginalis, and Neisseria gonorrhoeae. *Human Pathology, 21*(8), 831–837. https://doi.org/10.1016/0046-8177(90)90052-7.

Koch, R., Challagundla, N., Stelzner, K., & Rudel, T. (2025). *Chlamydia trachomatis* subverts neutrophil cell death pathways through RIP3 and Mcl-1 manipulation. *mBio, 16*, e02098-25. https://doi.org/10.1128/mbio.02098-25.

Kuang, Y., Parthasarathy, U., & Martinelli, R. (2023). Protocol for density gradient neutrophil isolation and flow cytometry-based characterization from human peripheral blood. *STAR Protocols, 4*(3), 102497. https://doi.org/10.1016/j.xpro.2023.102497.

Kuijpers, T. W., Tool, A. T. J., van der Schoot, C. E., Ginsel, L. A., Onderwater, J. J. M., Roos, D., & Verhoeven, A. J. (1991). Membrane surface antigen expression on neutrophils: A reappraisal of the use of surface markers for neutrophil activation. *Blood, 78*(4), 1105–1111. https://doi.org/10.1182/blood.V78.4.1105.1105.

Lakschevitz, F. S., Hassanpour, S., Rubin, A., Fine, N., Sun, C., & Glogauer, M. (2016). Identification of neutrophil surface marker changes in health and inflammation using high-throughput screening flow cytometry. *Experimental Cell Research, 342*(2), 200–209. https://doi.org/10.1016/j.yexcr.2016.03.007.

Lane, T. A., Bergum, P. W., Lichter, J. P., & Spragg, R. G. (1982). The labeling of rabbit neutrophils with [111In]oxine. *Journal of Immunological Methods, 51*(3), 293–305. https://doi.org/10.1016/0022-1759(82)90396-9.

Laster, S. M., Wood, J. G., & Gooding, L. R. (1988). Tumor necrosis factor can induce both apoptic and necrotic forms of cell lysis. *Journal of Immunology, 141*(8), 2629–2634.

Macey, M. G., McCarthy, D. A., Vordermeier, S., Newland, A. C., & Brown, K. A. (1995). Effects of cell purification methods on CD11b and l-selectin expression as well as the adherence and activation of leucocytes. *Journal of Immunological Methods, 181*(2), 211–219. https://doi.org/10.1016/0022-1759(95)00003-S.

Mastellos, D. C., Hajishengallis, G., & Lambris, J. D. (2024). A guide to complement biology, pathology and therapeutic opportunity. *Nature Reviews. Immunology, 24*(2), 118–141. https://doi.org/10.1038/s41577-023-00926-1.

Mayadas, T. N., Cullere, X., & Lowell, C. A. (2014). The multifaceted functions of neutrophils. *Annual Review of Pathology, 9*, 181–218. https://doi.org/10.1146/annurev-pathol-020712-164023.

McKenna, E., Mhaonaigh, A. U., Wubben, R., Dwivedi, A., Hurley, T., Kelly, L. A., ... Molloy, E. J. (2021). Neutrophils: Need for standardized nomenclature [review]. *Frontiers in Immunology, 12, 2021*. https://doi.org/10.3389/fimmu.2021.602963.

Merah-Mourah, F., Cohen, S. O., Charron, D., Mooney, N., & Haziot, A. (2020). Identification of novel human monocyte subsets and evidence for phenotypic groups defined by interindividual variations of expression of adhesion molecules. *Scientific Reports, 10*(1), 4397. https://doi.org/10.1038/s41598-020-61022-1.

Moore, P. L., Bank, H. L., Brissie, N. T., & Spicer, S. S. (1978). Phagocytosis of bacteria by polymorphonuclear leukocytes. A freeze-fracture, scanning electron microscope, and thin-section investigation of membrane structure. *The Journal of Cell Biology, 76*(1), 158–174. https://doi.org/10.1083/jcb.76.1.158.

Newton, K., Strasser, A., Kayagaki, N., & Dixit, V. M. (2024). Cell death. *Cell, 187*(2), 235–256. https://doi.org/10.1016/j.cell.2023.11.044.

Niu, H., Kozjak-Pavlovic, V., Rudel, T., & Rikihisa, Y. (2010). Anaplasma phagocytophilum Ats-1 Is imported into host cell mitochondria and interferes with apoptosis induction. *PLoS Pathogens, 6*(2), e1000774. https://doi.org/10.1371/journal.ppat.1000774.

Ovadia, S., Özcan, A., & Hidalgo, A. (2023). The circadian neutrophil, inside-out. *Journal of Leukocyte Biology, 113*(6), 555–566. https://doi.org/10.1093/jleuko/qiad038.

Papayannopoulos, V., Metzler, K. D., Hakkim, A., & Zychlinsky, A. (2010). Neutrophil elastase and myeloperoxidase regulate the formation of neutrophil extracellular traps. *The Journal of Cell Biology, 191*(3), 677–691. https://doi.org/10.1083/jcb.201006052.

Patton, D. L., & Kuo, C.-C. (1989). Histopathology of Chlamydia trachomatis salpingitis after primary and repeated reinfections in the monkey subcutaneous pocket model. *Reproduction (Cambridge, England), 85*(2), 647–656. https://doi.org/10.1530/jrf.0.0850647.

Pillay, J., den Braber, I., Vrisekoop, N., Kwast, L. M., de Boer, R. J., Borghans, J. A., ... Koenderman, L. (2010). In vivo labeling with 2H2O reveals a human neutrophil lifespan of 5.4 days. *Blood, 116*(4), 625–627. https://doi.org/10.1182/blood-2010-01-259028.

Rajeeve, K., Das, S., Prusty, B. K., & Rudel, T. (2018). Chlamydia trachomatis paralyses neutrophils to evade the host innate immune response. *Nat Microbiol, 3*(7), 824–835. https://doi.org/10.1038/s41564-018-0182-y.

Saba, S., Soong, G., Greenberg, S., & Prince, A. (2002). Bacterial stimulation of epithelial G-CSF and GM-CSF expression promotes PMN survival in CF airways. *American Journal of Respiratory Cell and Molecular Biology, 27*(5), 561–567. https://doi.org/10.1165/rcmb.2002-0019OC.

Savill, J., Dransfield, I., Gregory, C., & Haslett, C. (2002). A blast from the past: Clearance of apoptotic cells regulates immune responses. *Nature Reviews. Immunology, 2*(12), 965–975. https://doi.org/10.1038/nri957.

Savill, J., & Fadok, V. (2000). Corpse clearance defines the meaning of cell death. *Nature, 407*(6805), 784–788. https://doi.org/10.1038/35037722.

Savill, J. S., Wyllie, A. H., Henson, J. E., Walport, M. J., Henson, P. M., & Haslett, C. (1989). Macrophage phagocytosis of aging neutrophils in inflammation. Programmed cell death in the neutrophil leads to its recognition by macrophages. *The Journal of Clinical Investigation, 83*(3), 865–875. https://doi.org/10.1172/JCI113970.

Schofield, C. J., Tirouvanziam, R., & Garratt, L. W. (2024). OMIP-100: A flow cytometry panel to investigate human neutrophil subsets. *Cytometry Part A, 105*(2), 81–87. https://doi.org/10.1002/cyto.a.24820.

Shi, J., Gilbert, G. E., Kokubo, Y., & Ohashi, T. (2001). Role of the liver in regulating numbers of circulating neutrophils. *Blood, 98*(4), 1226–1230. https://doi.org/10.1182/blood.v98.4.1226.

Simon, H. U. (2003). Neutrophil apoptosis pathways and their modifications in inflammation. *Immunological Reviews, 193*, 101–110. https://doi.org/10.1034/j.1600-065x.2003.00038.x.
Tofts, P. S., Chevassut, T., Cutajar, M., Dowell, N. G., & Peters, A. M. (2011). Doubts concerning the recently reported human neutrophil lifespan of 5.4 days. *Blood, 117*(22), 6050–6052. author reply 6053-6054. https://doi.org/10.1182/blood-2010-10-310532.
Urban, C. F., Ermert, D., Schmid, M., Abu-Abed, U., Goosmann, C., Nacken, W., ... Zychlinsky, A. (2009). Neutrophil extracellular traps contain calprotectin, a cytosolic protein complex involved in host defense against Candida albicans. *PLoS Pathogens, 5*(10), e1000639. https://doi.org/10.1371/journal.ppat.1000639.
Wagner, F. Figure 7.1C: Blood layers. Created in BioRender. https://BioRender.com/wemyw6z.
Walker, R. I., & Willemze, R. (1980). Neutrophil kinetics and the regulation of granulopoiesis. *Reviews of Infectious Diseases, 2*(2), 282–292. http://www.jstor.org/stable/4452434.
Wang, J., Lu, S., Zheng, K., He, Z., Li, W., Liu, J., ... Wu, Y. (2022). Treponema pallidum delays the apoptosis of human polymorphonuclear neutrophils through the intrinsic and extrinsic pathways. *Molecular Immunology, 147*, 157–169. https://doi.org/10.1016/j.molimm.2022.04.012.
Wang, X., Yousefi, S., & Simon, H.-U. (2018). Necroptosis and neutrophil-associated disorders. *Cell Death & Disease, 9*(2), 111. https://doi.org/10.1038/s41419-017-0058-8.
Watson, R. W., Rotstein, O. D., Nathens, A. B., Parodo, J., & Marshall, J. C. (1997). Neutrophil apoptosis is modulated by endothelial transmigration and adhesion molecule engagement. *The Journal of Immunology, 158*(2), 945–953. https://doi.org/10.4049/jimmunol.158.2.945.
Whitlock, B. B., Gardai, S., Fadok, V., Bratton, D., & Henson, P. M. (2000). Differential roles for αMβ2 integrin clustering or activation in the control of apoptosis via regulation of Akt and ERK survival mechanisms. *Journal of Cell Biology, 151*(6), 1305–1320. https://doi.org/10.1083/jcb.151.6.1305.
Winterbourn, C. C., Hampton, M. B., Livesey, J. H., & Kettle, A. J. (2006). Modeling the reactions of superoxide and myeloperoxidase in the neutrophil phagosome: Implications for microbial killing. *The Journal of Biological Chemistry, 281*(52), 39860–39869. https://doi.org/10.1074/jbc.M605898200.
Youssef, P. P., Mantzioris, B. X., Roberts-Thomson, P. J., Ahern, M. J., & Smith, M. D. (1995). Effects of ex vivo manipulation on the expression of cell adhesion molecules on neutrophils. *Journal of Immunological Methods, 186*(2), 217–224. https://doi.org/10.1016/0022-1759(95)00146-2.
Zhang, F., Xia, Y., Su, J., Quan, F., Zhou, H., Li, Q., ... Jiang, Z. (2024). Neutrophil diversity and function in health and disease. *Signal Transduction and Targeted Therapy, 9*(1), 343. https://doi.org/10.1038/s41392-024-02049-y.

CHAPTER EIGHT

Real-time tracking of cell death in human intestinal organoids: A quantitative imaging approach

Jiyi Pang[a,b,*], Hojin Chang[c], Lachlan Whitehead[a,b], and James E. Vince[a,b,*]

[a]The Walter and Eliza Hall Institute of Medical Research, Parkville, Australia
[b]Department of Medical Biology, University of Melbourne, Parkville, Australia
[c]Department of Biochemistry & Pharmacology & Biological Optical Microscopy Platform, The University of Melbourne, Parkville, VIC, Australia
*Corresponding authors. e-mail address: pang.ji@wehi.edu.au; vince@wehi.edu.au

Contents

Abstract

Organoids are self-renewing three-dimensional tissue models that can be derived from patient samples. Their recapitulation of the structure and function of native tissues makes them powerful tools for studying development, physiology, disease mechanisms and personalized medicine in vitro. However, quantitative assessment of cell death in organoid systems remains challenging due to their structural complexity and dynamic responses. Here, we present a robust pipeline combining real-time

Methods in Cell Biology, Volume 210
ISSN 0091-679X, https://doi.org/10.1016/bs.mcb.2026.05.004

Incucyte® live-cell imaging with ImageJ-based quantitative analysis to measure cell death kinetics in human intestinal organoids. Organoids are cultured in the low-viscosity matrix suspension culture method, labelled with cell death–specific Incucyte® Cytotox Red Dye, and imaged via brightfield and fluorescence channels. A custom ImageJ analysis pipeline enables calculation of cell death overtime with high sensitivity and reproducibility. This approach overcomes the limitations of endpoint assays, delivering precise kinetic quantification of organoid cell death in response to cytokine stimulation, drug treatments, or genetic perturbations. The workflow is broadly applicable across diverse organoid systems and offers a scalable, standardized platform for interrogating cell death dynamics in basic and translational research.

1. Introduction

Organoid technology has rapidly evolved over the past 15 years. In 2009, Sato et al. reported that mouse intestinal *Lgr5*$^+$ stem cells can self-organize into crypt–villus structures in vitro (Sato et al., 2009), and published their development of human small intestine and colon organoids in 2011 (Sato et al., 2011). Since then, organoid models have been established for nearly every major organ system—including brain (Lancaster et al., 2013), liver (Huch et al., 2013), kidney (Takasato et al., 2015), lung (Dye et al., 2015), and tumor tissue, providing opportunities to study organ development (Kretzschmar & Clevers, 2016), host–pathogen interactions (Co et al., 2019; Noel et al., 2017), and patient-specific disease mechanisms (Wang et al., 2022). Moreover, organoids derived from patients retain phenotypic and genetic stability, closely mirroring the genome of their primary tissues. Therefore, organoids are increasingly used in high-throughput drug screening (Tan et al., 2024; Tan et al., 2023), regenerative medicine (Watanabe et al., 2022), and precision oncology (Tong et al., 2024), bridging the gap between traditional cell lines and animal models.

Organoids can be harnessed to provide valuable insights into cell death responses in physiological, pathological and therapeutic processes. Dysregulated programmed cell death, including apoptosis, necroptosis and pyroptosis, is involved in many pathological conditions (Anderton et al., 2020; Lawlor et al., 2024; Wallach et al., 2014). For example, excessive cell death has been linked to inflammatory bowel disease (IBD) (Blander, 2018; Patankar & Becker, 2020), SARS-CoV-2 (Junqueira et al., 2022; Karki et al., 2021; Liang et al., 2024; Rodrigues et al., 2021; Sefik et al., 2022; Simpson et al., 2022) and cytokine storm syndromes (Shojaee et al., 2025). Organoid systems provide a physiologically relevant 3D context that captures the spatial organization and cellular diversity of native tissues, enabling detailed dissection of programmed cell death pathways in settings often inaccessible in 2D culture. Moreover, rare or specialized cell types that are underrepresented in other models can be enriched in organoids, opening new opportunities to study their regulation (Boonekamp et al., 2020; Takashima et al., 2019). Recent work has applied organoid systems to explore viral cytopathology, including SARS-CoV-2 infection (Lamers et al., 2020), and to investigate genome instability-driven or inflammatory cytokine-triggered cell death signaling in gastrointestinal disorders (Flood et al., 2022; Wang et al., 2020). On the other hand, tumor-derived organoids have emerged as powerful translational platforms for testing chemotherapy efficacy, with cytotoxicity being one of the most common readouts (Driehuis et al., 2020). This underscores the utility of organoid-based cell death assays across both basic and applied research domains.

Despite the growing importance of cell death as a readout in organoid research, current methods to quantify this process, including LDH release assays, ATP-based assays (Driehuis et al., 2020), flow cytometry (Philippi et al., 2024) and immunostaining (Bode et al., 2019; Skorda et al., 2023), present significant limitations. These methods are generally endpoint, low-throughput, or technically confounded by organoid complexity. Thus, there remains a need for standardized, real-time strategies to measure cell death in 3D organoid systems with temporal resolution and single-organoid granularity. Here, we establish a workflow combining Incucyte® live-cell imaging with ImageJ-based analysis to quantify cell death in organoids cultured using a low-viscosity matrix suspension method (Hirokawa et al., 2021; Tan et al., 2022), enabling dynamic, quantitative, and scalable assessment of organoid death.

2. Materials

2.1 Common disposables

- Falcon® 24-well Polystyrene Clear Flat Bottom Not Treated Cell Culture Plate, with Lid, Individually Wrapped, Sterile (Corning, 351147)
- Falcon® 96-well Clear Flat Bottom Not Treated Cell Culture Microplate, with Lid, Individually Wrapped, Sterile (Corning, 351172)
- Conical Tube 10 mL, 100 × 16 PP + Cap (SARSTEDT, 62.9924.284)
- Nunc™ 50 mL Conical Sterile Polypropylene Centrifuge Tubes (Thermo Fisher Scientific, 339652),

- Terumo Syringe Luer Slip Tip – 1cc/1 mL
- BD Microlance™3 Needles Sterile, 25 G 25 mm Regular Orange (300600)
- Eppendorf™ 1.5 mL Flex-Tube™ - Microtube (Eppendorf, 022364111)

2.2 Reagents

- IntestiCult™ Organoid Growth Medium (Human) (OGM, STEMCELL Technologies, 06010) *(see Note 1)*
- IntestiCult™ Organoid Differentiation Medium (Human) (ODM, STEMCELL Technologies, 100-0214) *(see Note 1)*
- IWP-2 (STEMCELL Technologies,72122) *(see Note 2)*
- Bovine Serum Albumin (BSA, Sigma-Aldrich, A4612)
- TrypLE Express enzyme (Thermo Fisher Scientific, 12604021)
- DMEM/F-12 with 15 mM HEPES (DMEM/F-12, STEMCELL Technologies, 36254)
- Corning® Matrigel® Basement Membrane Matrix, Phenol Red-free, LDEV-free, 10 mL (Bio-strategy, 356237)
- Primocin® (Invivogen, ant-pm-05)
- Gentamicin (Thermo Fisher Scientific, 15710064)
- EDTA chelation buffer (Sigma-Aldrich, E5134)
- Dithiothreitol (Merck, 10197777001)
- Penicillin-streptomycin (Life Technologies, 15140122)
- Rho-kinase inhibitor Y27632 (Stemcell Technologies, 72308)
- Incucyte® Cytotox Red Dye (Sartorius, 4632) *(see Note 3)*
- Phosphate-buffered saline (PBS) *(see Note 4)*
- IFNγ (recombinant human, R&D, 285-IF-100)
- Human Fc-TNF (in-house or Enzo, ALX-522-087)
- ABT-737 (Active Biochem; A-6044)
- Cycloheximide (CHX, Sigma; C7698)

2.3 Equipment

- Laboratory biosafety cabinet (Class II), such as Purifier Logic + Class II Type A2 Biological Safety Cabinet (#320419101, Labconco)
- Humidified cell culture incubator, such as CelCulture® CO_2 Incubator (Esco lifescience)
- Benchtop centrifuge, such as Sorvall™ Legend™ Micro 17 Microcentrifuge (Thermo fisher)
- Floor-standing centrifuge, such as Thermo Scientific Megafuge 40R Refrigerated Centrifuge

- Hemocytometer such as Neubauer or a Burker chamber.
- IncuCyte S3 Live-Cell Analysis System (Sartorius)

2.4 Software

- ImageJ 1.53t
- Excel (Microsoft)
- Prism v. 10.0.1 (GraphPad)

3. Methods

3.1 Organoid generation and maintenance

1. Prepare full organoid growth medium by supplementing OGM with 100 U/mL penicillin-streptomycin, 10 μM Y-27632 and 20 μg/mL Gentamicin. Mix well and place on ice *(see Note 5).*
2. Prepare DMEM/F-12+ 1 % BSA medium by mixing 10 % w/v BSA stock with DMEM/F-12 +15 mM HEPES (e.g., 10 mL 10 % BSA + 90 mL DMEM/F-12 with HEPES). Mix well by inversion. Place on ice. *(see Note 6)*
3. Thaw Matrigel on ice. *(see Note 7)*
4. Collect fresh human intestinal specimens and place specimens in DMEM/F-12 with 15 mM HEPES media supplemented with 1 % w/v BSA on ice until processing.
5. Transfer tissue into a 10 mL conical tube. Wash the tissue sample with 5 mL of ice-cold PBS containing 100 μg/mL Primocin® and 20 μg/mL Gentamicin. Allow the tissue to settle by gravity (~ 5 s), then aspirate supernatant carefully. *(see Note 8)*
6. Perform 2 additional washes using the same PBS solution.
7. Incubate tissue in 1 mL of 3 mM EDTA chelation buffer containing 100 μM DTT. Keep at room temperature for ~30 min, shaking vigorously every few minutes to facilitate crypt release. *(see Note 9)*
8. Remove the remaining tissue, collect the crypts by centrifuge at 500 × *g* for 3 min at 4 °C.
9. Discard supernatant, resuspend crypts in 200 μL DMEM/F-12 + 1 % BSA.
10. Mix crypts suspension with 3 mL of supplemented OGM. Plate 1 mL/well into sterile, non-tissue culture treated 24-well plates (Fig. 1A). *(see Note 10)*
11. Add 5 % v/v Matrigel directly to each well, mix gently by pipetting up and down.

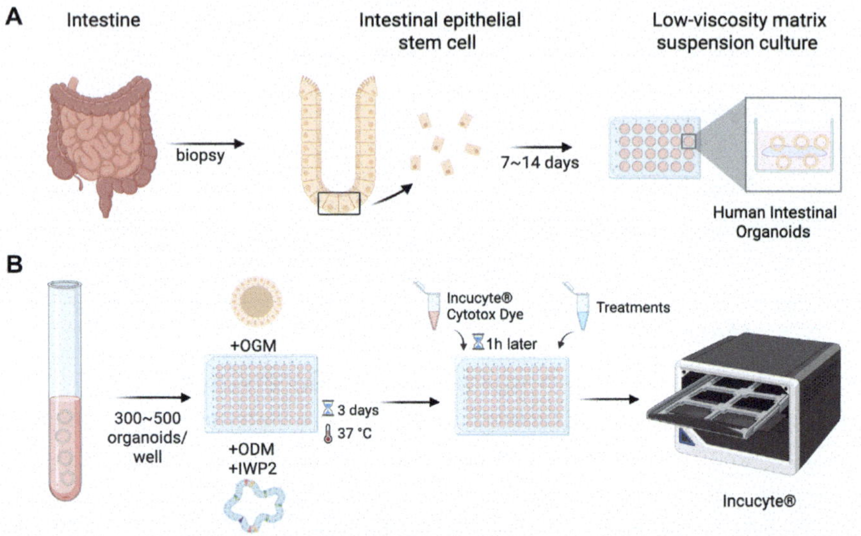

Fig. 1 Schematic for organoid generation and measuring cell death. (A) Organoids are derived from intestinal biopsies and maintained in low-viscosity matrix culture. (B) Organoids are plated in 96-well plates and differentiated using distinct growth conditions for Incucyte imaging-based cell death analysis.

12. Incubate plates at 37 °C in 5 % CO_2. Refresh medium every 2 days by adding 200 μL fresh OGM per well.
13. Maintain cultures under the same conditions for 7–14 days (Fig. 1A).

3.2 Organoids passaging, plating and differentiation

1. Thaw Matrigel on ice. Place DMEM + 1 % BSA and PBS on ice.
2. Collect organoids into a 10 mL conical tube by gently pipetting up and down to break Matrigel.
3. Wash the empty well once with pre-chilled PBS and combine the wash with the collected organoid suspension.
4. Centrifuge at 300 × *g* for 3 min at 4 °C. Discard the supernatant.
5. Wash organoids one time by resuspending pellet in 5 mL of cold PBS, centrifuge again at 300 × *g* for 3 min at 4 °C. Discard the supernatant.
6. Resuspend organoids in 1 mL TrypLE Express and transfer to a 1.5 mL microcentrifuge tube. *(see Note 11)*
7. Incubate at 37 °C for 10 min, shaking gently every few minutes to facilitate digestion. *(see Note 12)*
8. After 10 min incubation, stop digestion by centrifuging cells at 10,000 rpm, 4 °C for 30 s. Discard the supernatant.

9. Add 1 mL of DMEM/F-12 + 1 % BSA to resuspend cells. Repeat centrifugation at 10,000 rpm, 4 °C for 30 s
10. Discard the supernatant and resuspend the pellet in 500 μL DMEM/F−12 + 1 % BSA.
11. Mechanically dissociate cells by passing through a 1 mL syringe with a 26G needle until a single-cell suspension is obtained. *(see Note 13)*
12. Count cells using a hemocytometer.
13. (24-well plate format) Seed 1×10^5 cells per well in 1 mL OGM supplemented with 100 U/mL penicillin-streptomycin, 10 μM Y-27632, and 5 % Matrigel. Mix gently and plate in sterile, non-tissue culture treated 24-well plates.
14. Incubate plates at 37 °C, 5 % CO_2 incubator. Refresh medium every 2 days by adding 200 μL fresh OGM per well.
15. Maintain cultures for 7–10 days, or re-plate organoids 4-~5 days post-passaging, as needed.
16. Organoids can be replated into non-tissue culture treated 96-well plates for differentiation and further treatments. *(see Note 14)*
17. For 96 well plating, day 4–5 organoids are collected in a 10 mL conical tube and washed with PBS once following Steps 2−5, Section 3. 2. *(see Note 15)*
18. Resuspend organoids in 1 mL DMEM/F−12 + 1 % BSA. Pipette organoids multiple times to ensure even distribution. *(see Note 15)*
19. Place 10 μL aliquots of organoid suspension on both sides of a hemocytometer. Count the number of organoids using 4x magnification. The total organoid number will be the average count x100.
20. 300–500 organoids can be plated in each well of a non-tissue culture treated 96-well plate in a volume of 50 μL *(see Note 16)*. Based on the number of wells required prepare organoids resuspension media as follow:
    a. For stem cell organoids: OGM + 5 % Matrigel
    b. For organoid differentiation: ODM + 5 % Matrigel ± small molecules. For example, we routinely add 10 μM IWP−2 to differentiate organoids into colonocytes.
21. Spin down organoids at $300 \times g$ for 3 min at 4 °C. Discard the supernatant. Resuspend organoid pellets in the appropriate media formulation from Step 20.
22. Plate organoids into non-tissue culture treated 96-well tissue culture plates. Gently shake the plate to evenly distribute organoids.
23. Fill surrounding empty wells with 200 uL PBS to prevent evaporation.

24. Place the plate in a 37 °C, 5 % CO_2 incubator.
25. For undifferentiated organoids, proceed with treatments after 1 day. For differentiated organoids, incubate for 3 nights before treatment (Fig. 1B).

3.3 Organoid treatment and Incucyte setup

1. Add Incucyte Cytotox Red Dye to organoids at a 1:20,000 dilution (Fig. 1B). *(see Note 17).*
2. Incubate organoids with Cytotox Red Dye for 0.5 h at 37 °C prior to other treatments.
3. Organoids are then treated with any required compounds. To avoid adding large liquid volumes into the organoid culture system, it is recommended that drug treatments are prepared by serial dilution from concentrated stocks. For cell death assay demonstration, three representative conditions were applied here: untreated control, IFNγ and TNF treatment, ABT-737 and cycloheximide treatment (Fig. 2A) *(see Note 18)*
4. Acquire images using the Incucyte system with the following settings:
   a. Objective: 4 × *(see Note 19, 20)*
   b. Scan type: Spheroid
   c. Channels: Phase/Brightfield (BF) and Red fluorescence
   d. Exposure time: 250 ms (Red channel)
5. Collect images throughout the assay at defined intervals. For assay demonstration, organoids treated with untreated control, IFNγ/TNF and ABT-737/cycloheximide were imaged over 24 h at 2-hour intervals (Fig. 2B).
6. After scanning, export both brightfield (BF) and red channel uncalibrated images into separate folders. *(see Note 21)*
7. Analyze images with ImageJ v1.53t using custom semi-automated macros.

3.4 Cell death analysis

1. Images acquired from the Incucyte are processed using FIJI/ImageJ (v1.53t) with two custom macros developed to enable reproducible, batch-based quantification of organoid death across large datasets (see the full script and explanation in the Appendix) *(see Note 22).*
2. To standardize the region of interest and minimize edge-associated illumination artefacts introduced by well boundaries, a rectangular crop to exclude edge artifacts is first defined on a representative BF image using the Recorder tool (Fig. 2A). *(see Note 23)*

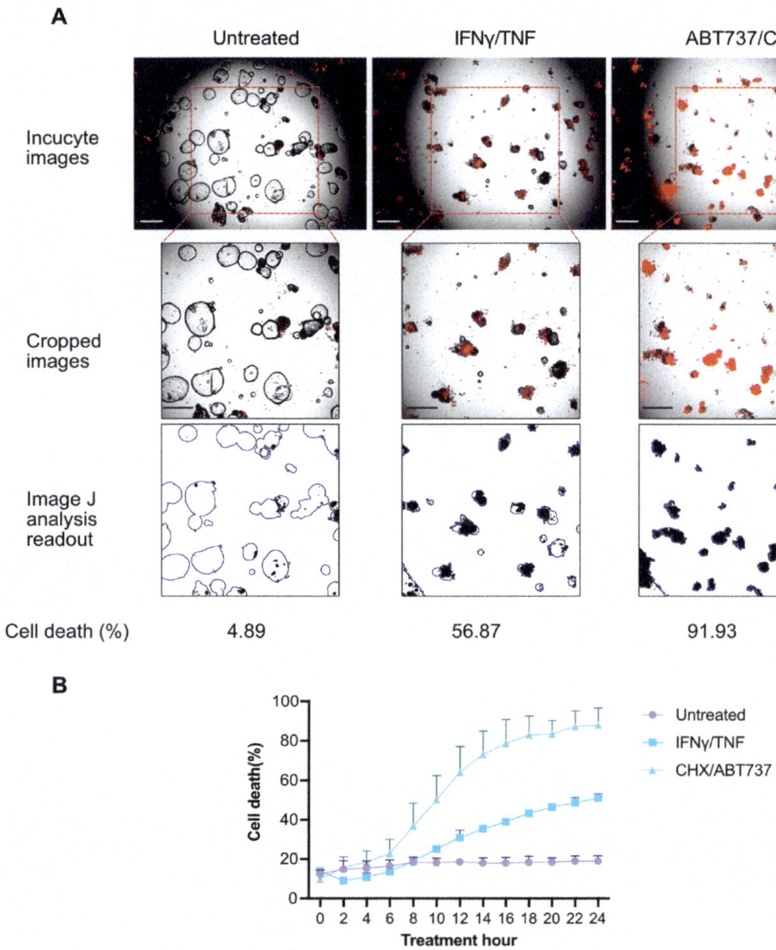

Fig. 2 Analysis of organoid death. (A) Images acquired from the Incucyte are cropped to remove edge effects then processed using FIJI/ImageJ (v1.53t) with two custom macros to define the total area of organoids and the area of organoids that are dead - positive for Cytotox Red Dye. Representative images where intestinal organoids were treated with IFNγ (50 ng/mL)/TNF (50 ng/mL) and ABT-737 (1 μM)/cycloheximide (20 μg/mL) for 24 h to cause cell death are shown. Scale bar = 500 μm. (B) Measurements (area, mean, min, max, and % red area) exported automatically to a custom results table are graphed to plot the % organoid death over time (% Cytotox Red positive area of brightfield, mean +SD of 3 triplicate wells).

3. The recorded crop parameters are then applied to all images using the Batch Process function, generating cropped images for both BF and red channels, and ensuring consistent spatial sampling across time points and conditions.

4. Save cropped BF/red channel images into output folders.
5. Merge cropped BF/red channel images into single stacked images using a custom ImageJ macro 1 (Appendix).
6. Run custom ImageJ macro 2 (Appendix) for the following processes:
   a. BF images are preprocessed with the MorphoLibJ plugin (Gray Scale Attribute Filtering) (Legland et al., 2016) to improve contrast and suppress background, enabling robust segmentation from BF.
   b. Then the images are converted into a binary mask using a fixed intensity threshold (0–132 pixel values). This threshold was established during initial pipeline optimization using representative BF images to robustly capture intact organoid structures while excluding background. Segmented objects smaller than 500 pixels are excluded to remove debris and non-organoid features. The remaining objects are combined to define a Region of Interest (ROI) representing the total organoid area. *(see Note 24)*
   c. The Cytotox Red channel is quantified exclusively within this BF-defined ROI to ensure that only cell death associated with organoids is measured.
   d. A manual/empirically optimized threshold (e.g., 140–65,535) is applied to the red channel. This thresholding approach is chosen to accommodate experiment-to-experiment variability in signal intensity while maintaining a consistent biological definition of death-associated fluorescence *(see Note 25)*. Threshold values should be held constant within individual experiments.
   e. Cell death is quantified as the percentage of red-positive area relative to the total organoid area.
   f. Measurements (area, mean, min, max, and % red area) are exported automatically to a custom results table for downstream analysis.

4. Notes

1. The organoid growth and differentiation media described here can be replaced with conditioned media as reported in previous studies, such as Sato *et al.* (Sato et al., 2011), Miyoshi et al. (Miyoshi & Stappenbeck, 2013), Fujii et al. (Fujii et al., 2018) and He et al. (He et al., 2022).
2. IWP-2 is used here as a Notch pathway inhibitor to promote organoid enterocyte/colonocyte differentiation (Yin et al., 2014). Other small

molecules may be substituted to direct differentiation toward specific intestinal lineages, depending on the experimental objective.

3. Cytotox Red Dye can be substituted with Propidium Iodide, Cytotox Green, or other fluorescent cell death markers, provided the fluorophore is compatible with the Incucyte imaging channels.
4. PBS can also be prepared according to the following recipe: 137 mM NaCl, 10 mM Na2HPO4, and 1.8 mM KH2PO4, pH: 7.4.
5. OGM and penicillin-streptomycin should be premixed, aliquoted into 10 mL conical tubes, and stored at −20 °C for up to 3 months. Before use, thaw the required volume and, once thawed, use within 1 week.
6. DMEM/F-12 supplemented with 1% BSA can be stored at 4 °C for up to 6 months.
7. The entire bottle of Matrigel should be thawed at 4 °C overnight and aliquoted into small vials (e.g., 500 μL per tube). Aliquots should be stored at −20 °C and thawed on ice ~30 min before use. Matrigel must be kept on ice throughout handling.
8. Use cut 1 mL pipette tips to facilitate the transfer of tissue.
9. During EDTA/DTT digestion check for crypt release approximately every 10 min. Stop the reaction once the majority of crypts are released to avoid over-digestion.
10. Organoids can also be cultured in non-tissue culture treated 6-well plates by adding 3 mL of supplemented OGM + 5 % v/v Matrigel per well.
11. Organoids are transferred into microcentrifuge tubes to facilitate the subsequent mechanical dissociation with a syringe needle.
12. Organoids can be incubated at 37 °C either in a standard incubator or on a 37 °C orbital shaker.
13. During mechanical dissociation, orient the beveled tip of the needle against the wall of the microcentrifuge tube to minimize bubble formation. Typically, 15–20 passes are sufficient to disrupt organoids into single cells. Excessive pipetting should be avoided to prevent mechanical stress. Dissociation efficiency can be verified by transferring a small aliquot to a 24-well plate and examining cells under a microscope.
14. 384-well plates are also compatible with Incucyte imaging and can be used as an alternative to 96-well plates.
15. To maintain the structural integrity of organoids, the end of pipette tips can be cut prior to pipetting.

16. For the 384-well format, seeding ~100 organoids per well in 30 μL of culture medium is recommended.
17. Incucyte Cytotox Red Dye can be pre-diluted 1:20 in DMSO, followed by dilution 1:1000 in culture medium to achieve the working concentration.
18. For each treatment condition, it is recommended to include 2–3 replicate wells of organoids to ensure reproducibility.
19. When using 384-well plates, a 10 × objective should be selected for optimal image resolution.
20. Labels should not be placed on the plate lid, as this can interfere with the Incucyte's automated focusing.
21. When using 384-well plates, only phase images can be captured and exported with the incucyte system.
22. Although developed using Incucyte images, this workflow is compatible with any imaging platform that outputs time-lapse brightfield and fluorescence images in standard formats (e.g., TIFF), and can be readily adapted by adjusting threshold parameters.
23. Cropping should be applied to exclude edge shadowing artifacts, as these can interfere with accurate organoid area quantification
24. For 384-well plates, phase images rather than brightfield images are converted into binary masks, using an intensity threshold of 0–126 pixel values. The subsequent analysis workflow remains the same as for brightfield images
25. The choice of threshold values is crucial for reproducibility. Users should optimize thresholds using representative images that capture the full dynamic range of the assay (e.g.,healthy organoids, intermediate death, and extensive death).

5. Concluding remarks

Quantifying cell death overtime in organoid systems remains a methodological challenge due to the structural complexity and heterogeneity of 3D cultures in addition to often limited amounts of sample material. Traditional assays such as LDH release, ATP depletion, flow cytometry, and immunostaining provide valuable insights but are largely limited to endpoint readouts, low throughput, or are technically confounded by organoid processing. In this Chapter, we introduce a live-cell imaging workflow that integrates Incucyte® time-lapse microscopy with ImageJ-based analysis, enabling continuous and quantitative monitoring of cell death dynamics at single-organoid resolution. This platform not only captures the kinetics of cell death but also provides morphological context, making it well suited for studies of cytokine responses, pharmacological screening, and genetic perturbation. Furthermore, its compatibility with 96- and 384-well plate formats facilitates scalability, establishing a standardized and versatile approach for the broader organoid research community. By bridging the gap between traditional endpoint assays and real-time analysis, this method offers a practical and adaptable tool for advancing cell death research in complex 3D systems.

At the same time, several caveats should be considered. The method is compatible with suspension-based formats, but dome-style Matrigel cultures can be challenging due to difficulties in focusing and increased background signal compared to suspension methods. These issues may be mitigated by reducing seeding density or applying more advanced image analysis strategies. The method also provides limited specificity for distinguishing between different death pathways and cannot fully resolve single-cell events in tightly packed organoids. Finally, careful thresholding in ImageJ/FIJI is required to distinguish true signal from background or debris, which can introduce variability. Despite these limitations, in our own research we have found that this method of cell death analysis is superior to other methodologies and can be used reproducibly to define human intestinal organoid cell death kinetics following the triggering of distinct cell death modalities.

Acknowledgments

This work was supported by a National Health and Medical Research Council of Australia (NHMRC) Investigator Grant (2008692) to James E Vince.

Code availability

All custom FIJI/ImageJ macros used for image processing and quantitative analysis are publicly available on Zenodo at 10.5281/zenodo.17938198, allowing direct download and implementation.

Conflicts of interests

The authors have no conflicts of interest to declare.

Appendix

Image J macros with explanations.

Macro 1 – Stack cropped images (BF + red channels into one stack).

```
// ===============================.
// FIJI Macro: Stack cropped BF+Red channel images.
```

```
// Purpose: takes paired cropped BF and Red channel images and stacks them into a single.tif file.
// Inputs: directory containing cropped BF and Red channel images (assumed in alternating order).
// Outputs: stacked.tif files saved to output_path.
// ===============================.

input_path = getDirectory("input files"); // Prompt user to select input folder.
fileList = getFileList(input_path); // Get list of all files in input folder.
output_path = " C:\\Users\\xxx\\"; // Set output folder (edit for your system).
for (f=0; f < fileList.length; f=f+2){ // Loop over files in pairs (BF + red channel).

    // Clean up ImageJ environment before each run.
    roiManager("reset");.
    run("Close All");.
    run("Clear Results");.
    // Open BF image.
    open(input_path + fileList[f]);.
    print(input_path + fileList[f]);.
    // Open corresponding Red channel image.
    open(input_path + fileList[f+1]);.
    print(input_path + fileList[f+1]);.
    // Use the second opened file as the stack title.
    Title = getTitle();.

    // Stack BF and Red images into one.tif.
    run("Images to Stack", "name=[title]");.
    rename(Title);.
    // Save stack to output folder.
    saveAs("TIF", output_path + Title + ".tif");.
    // Close all open images before next loop.
    run("Close All");.
}.
```

Macro 2 – Thresholding, segmentation, and quantification.

```
// ===============================.
// FIJI Macro: Organoid cell death quantification.
// Purpose: segment BF (organoid area), apply ROI to Red channel, quantify Cytotox Red-positive area.
// Inputs: stacked.tif images containing BF + Red channels.
// Outputs: segmented mask images and custom results table (per-image metrics).
// Adjustable: thresholds, input/output directories.
// ===============================.
```

```
dir1 = " C:\\Users\\xxx\\"; // Input folder (edit for your system).
flist = getFileList(dir1);.

// Setup custom results table with desired columns.
Table_Heading = "Table name";.
columns = "Filename,Area,Mean,Min,Max,Red % Area"; // Define measurement
columns.
columns = split(columns,",");.
table = generateTable(Table_Heading,columns);.

// Loop over all.tif stack files.
for(i = 0;i < flist.length;i + +){.
    if(endsWith(flist[i],"tif")){.
       open(dir1 +flist[i]); // Open stack.
       print("Processing: " + dir1 +flist[i]);.
       run("Stack to Images"); // Split stack into separate BF and Red images.
       BF = getTitle();.
       rename(BF + " BF"); // Rename first image as BF.
       selectWindow(BF);.
       rename(BF + " Red"); // Rename second image as Red channel.

       // —— BF channel processing ——
       selectWindow(BF + " BF");.
       // Remove small noise / smooth objects using MorphoLibJ filter.
       run("Gray Scale Attribute Filtering", "operation=Opening
       attribute=Area minimum=10000 connectivity=4");.
       run("Overlay Options...", "stroke=blue width=3 fill=none set show");.

       // Apply threshold to BF image to define organoid area (ROI).
       setAutoThreshold("Default stack");.
       setThreshold(0, 132); // Adjustable threshold for BF segmentation.
       setOption("BlackBackground", false);.
       run("Convert to Mask");.
       run("Analyze Particles...", "size = 500-Infinity pixel show=Outlines
       clear include overlay add");.
       n = roiManager("count"); // Count ROIs detected.
       run("Select All");.
       roiManager("Combine"); // Combine all ROIs into single ROI (total
       organoid area).
       roiManager("Add");.

       // Save BF mask output for QC.
       selectWindow(BF + " BF");.
       roiManager("select", n);.
       run("Add Selection...");.
       output_path = "C:\\Users\\xxx\\"; // Output folder (edit for your
       system).
       saveAs("TIF", output_path + BF + "_BF_mask.tif");.

       // —— Red channel processing ——
```

```
        selectWindow(BF + " Red");.
        roiManager("select", n); // Apply BF ROI to Red channel.
        run("Clear Outside"); // Restrict analysis to organoid area.
        setAutoThreshold("Default dark");.
        setThreshold(xx, 65535, "raw"); // Adjustable Red channel threshold.
        setOption("BlackBackground", false);.
        run("Convert to Mask");.

        // Measure Red-positive area within BF ROI.
        roiManager("select", n);.
        run("Set Measurements…", "area mean min area_fraction display
        redirect=None decimal=6");.
        run("Measure");.

        // Extract results for this image.
        fname = BF;.
        area = getResult("Area");.
        mean = getResult("Mean");.
        min = getResult("Min");.
        max = getResult("Max");.
        pc_area = getResult("%Area");.

        res = newArray(fname,area,mean,min,max,pc_area);.
        logResults(table,res); // Append results to custom table.

        // Save Red mask output for QC.
        roiManager("select", n);.
        run("Add Selection…");.
        output_path = "C:\\Users\\xxx\\"; // Output folder (edit for your
        system).
        saveAs("TIF", output_path + BF + "_Red_mask.tif");.

        // Clean up for next loop.
        roiManager("Deselect");.
        roiManager("Delete");.
        close("*");.
    }.
}.

// ============================.
// Helper functions for results table.
// ============================.

// Create a custom results table.
function generateTable(tableName,column_headings){.
    if(isOpen(tableName)){.
        selectWindow(tableName);.
        run("Close");.
    }.
    tableTitle=tableName;.
    tableTitle2 = "[" + tableTitle + "]";.
```

```
    run("Table...","name="+tableTitle2+" width=600 height=250");.
    newstring = "\\Headings:"+column_headings[0];.
    for(i=1;i < column_headings.length;i++){.
        newstring=newstring+" \t "+column_headings[i];.
    }.
    print(tableTitle2,newstring);.
    return tableTitle2;.
}.

// Log results into custom table.
function logResults(tablename,results_array){.
    resultString=results_array[0]; // First column=filename.
    for(i=1;i < results_array.length;i++){.
        resultString=toString(resultString + " \t " + results_array[i]);.
    }.
    print(tablename,resultString);.
}.
```

References

Anderton, H., Wicks, I. P., & Silke, J. (2020). Cell death in chronic inflammation: Breaking the cycle to treat rheumatic disease. *Nature Reviews Rheumatology, 16*(9), 496–513. https://doi.org/10.1038/s41584-020-0455-8.

Blander, J. M. (2018). On cell death in the intestinal epithelium and its impact on gut homeostasis. *Current Opinion in Gastroenterology, 34*(6), 413–419. https://doi.org/10.1097/MOG.0000000000000481.

Bode, K., Mueller, S., Schweinlin, M., Metzger, M., & Brunner, T. (2019). A fast and simple fluorometric method to detect cell death in 3D intestinal organoids. *Biotechniques, 67*. https://doi.org/10.2144/btn-2019-0023.

Boonekamp, K. E., Dayton, T. L., & Clevers, H. (2020). Intestinal organoids as tools for enriching and studying specific and rare cell types: Advances and future directions. *Journal of Molecular Cell Biology, 12*(8), 562–568. https://doi.org/10.1093/jmcb/mjaa034.

Co, J. Y., Margalef-Català, M., Li, X., Mah, A. T., Kuo, C. J., Monack, D. M., & Amieva, M. R. (2019). Controlling epithelial polarity: A human enteroid model for host-pathogen interactions. *Cell Reports, 26*(9), 2509–2520.e2504. https://doi.org/10.1016/j.celrep.2019.01.108.

Driehuis, E., Kretzschmar, K., & Clevers, H. (2020). Establishment of patient-derived cancer organoids for drug-screening applications. *Nature Protocols, 15*(10), 3380–3409. https://doi.org/10.1038/s41596-020-0379-4.

Dye, B. R., Hill, D. R., Ferguson, M. A., Tsai, Y. H., Nagy, M. S., Dyal, R., ... Spence, J. R. (2015). In vitro generation of human pluripotent stem cell derived lung organoids. *Elife, 4*. https://doi.org/10.7554/eLife.05098.

Flood, P., Fanning, A., Woznicki, J. A., Crowley, T., Christopher, A., Vaccaro, A., ... Nally, K. (2022). DNA sensor-associated type I interferon signaling is increased in ulcerative colitis and induces JAK-dependent inflammatory cell death in colonic organoids. *American Journal of Physiology-Gastrointestinal and Liver Physiology, 323*(5), G439–G460. https://doi.org/10.1152/ajpgi.00104.2022.

Fujii, M., Matano, M., Toshimitsu, K., Takano, A., Mikami, Y., Nishikori, S., ... Sato, T. (2018). Human intestinal organoids maintain self-renewal capacity and cellular diversity in niche-inspired culture condition. *Cell Stem Cell, 23*(6), 787–793.e786. https://doi.org/10.1016/j.stem.2018.11.016.

He, G. W., Lin, L., DeMartino, J., Zheng, X., Staliarova, N., Dayton, T., ... Clevers, H. (2022). Optimized human intestinal organoid model reveals interleukin-22-dependency of paneth cell formation. *Cell Stem Cell, 29*(9), 1333–1345.e1336. https://doi.org/10.1016/j.stem.2022.08.002.

Hirokawa, Y., Clarke, J., Palmieri, M., Tan, T., Mouradov, D., Li, S., ... Sieber, O. M. (2021). Low-viscosity matrix suspension culture enables scalable analysis of patient-derived organoids and tumoroids from the large intestine. *Communications Biology, 4*(1), 1067. https://doi.org/10.1038/s42003-021-02607-y.

Huch, M., Dorrell, C., Boj, S. F., van Es, J. H., Li, V. S., van de Wetering, M., ... Clevers, H. (2013). In vitro expansion of single Lgr5+ liver stem cells induced by Wnt-driven regeneration. *Nature, 494*(7436), 247–250. https://doi.org/10.1038/nature11826.

Junqueira, C., Crespo, Â., Ranjbar, S., de Lacerda, L. B., Lewandrowski, M., Ingber, J., ... Lieberman, J. (2022). FcγR-mediated SARS-CoV-2 infection of monocytes activates inflammation. *Nature, 606*(7914), 576–584. https://doi.org/10.1038/s41586-022-04702-4.

Karki, R., Sharma, B. R., Tuladhar, S., Williams, E. P., Zalduondo, L., Samir, P., ... Kanneganti, T. D. (2021). Synergism of TNF-alpha and IFN-gamma triggers inflammatory cell death, tissue damage, and mortality in SARS-CoV-2 infection and cytokine shock syndromes. *Cell, 184*(1), 149–168.e117. https://doi.org/10.1016/j.cell.2020.11.025.

Kretzschmar, K., & Clevers, H. (2016). Organoids: Modeling development and the stem cell niche in a dish. *Developmental Cell, 38*(6), 590–600. https://doi.org/10.1016/j.devcel.2016.08.014.

Lamers, M. M., Beumer, J., van der Vaart, J., Knoops, K., Puschhof, J., Breugem, T. I., ... Clevers, H. (2020). SARS-CoV-2 productively infects human gut enterocytes. *Science (New York, N. Y.), 369*(6499), 50–54. https://doi.org/10.1126/science.abc1669.

Lancaster, M. A., Renner, M., Martin, C. A., Wenzel, D., Bicknell, L. S., Hurles, M. E., ... Knoblich, J. A. (2013). Cerebral organoids model human brain development and microcephaly. *Nature, 501*(7467), 373–379. https://doi.org/10.1038/nature12517.

Lawlor, K. E., Murphy, J. M., & Vince, J. E. (2024). Gasdermin and MLKL necrotic cell death effectors: Signaling and diseases. *Immunity, 57*(3), 429–445. https://doi.org/10.1016/j.immuni.2024.02.011.

Legland, D., Arganda-Carreras, I., & Andrey, P. (2016). MorphoLibJ: Integrated library and plugins for mathematical morphology with ImageJ. *Bioinformatics (Oxford, England), 32*(22), 3532–3534. https://doi.org/10.1093/bioinformatics/btw413.

Liang, K., Barnett, K. C., Hsu, M., Chou, W. C., Bais, S. S., Riebe, K., ... Ting, J. P. (2024). Initiator cell death event induced by SARS-CoV-2 in the human airway epithelium. *Science Immunology, 9*(97), eadn0178. https://doi.org/10.1126/sciimmunol.adn0178.

Miyoshi, H., & Stappenbeck, T. S. (2013). In vitro expansion and genetic modification of gastrointestinal stem cells in spheroid culture. *Nature Protocols, 8*(12), 2471–2482. https://doi.org/10.1038/nprot.2013.153.

Noel, G., Baetz, N. W., Staab, J. F., Donowitz, M., Kovbasnjuk, O., Pasetti, M. F., & Zachos, N. C. (2017). A primary human macrophage-enteroid co-culture model to investigate mucosal gut physiology and host-pathogen interactions. *Scientific Reports,* 7(1), 45270. https://doi.org/10.1038/srep45270.

Patankar, J. V., & Becker, C. (2020). Cell death in the gut epithelium and implications for chronic inflammation. *Nature Reviews Gastroenterology & Hepatology, 17*(9), 543–556. https://doi.org/10.1038/s41575-020-0326-4.

Philippi, C. I., Hagens, J., Heuer, K. M., Schmidt, H. C., Schuppert, P., Pagerols Raluy, L., ... Tomuschat, C. (2024). Exploring cell death mechanisms in spheroid cultures using a novel application of the RIP3-caspase3-assay. *Scientific Reports, 14*(1), 16032. https://doi.org/10.1038/s41598-024-66805-4.

Rodrigues, T. S., de Sa, K. S. G., Ishimoto, A. Y., Becerra, A., Oliveira, S., Almeida, L., ... Zamboni, D. S. (2021). Inflammasomes are activated in response to SARS-CoV-2 infection and are associated with COVID-19 severity in patients. *The Journal of Experimental Medicine, 218*(3), https://doi.org/10.1084/jem.20201707.

Sato, T., Stange, D. E., Ferrante, M., Vries, R. G., Van Es, J. H., Van den Brink, S., ... Clevers, H. (2011). Long-term expansion of epithelial organoids from human colon, adenoma, adenocarcinoma, and Barrett's epithelium. *Gastroenterology, 141*(5), 1762–1772. https://doi.org/10.1053/j.gastro.2011.07.050.

Sato, T., Vries, R. G., Snippert, H. J., van de Wetering, M., Barker, N., Stange, D. E., ... Clevers, H. (2009). Single Lgr5 stem cells build crypt-villus structures in vitro without a mesenchymal niche. *Nature, 459*(7244), 262–265. https://doi.org/10.1038/nature07935.

Sefik, E., Qu, R., Junqueira, C., Kaffe, E., Mirza, H., Zhao, J., ... Flavell, R. A. (2022). Inflammasome activation in infected macrophages drives COVID-19 pathology. *Nature, 606*(7914), 585–593. https://doi.org/10.1038/s41586-022-04802-1.

Shojaee, F., Azadian, E., Wong, M. X., Ma, X., Rickard, J., Pang, J., ... Vince, J. E. (2025). NLRP3 inflammasome-driven hemophagocytic lymphohistiocytosis occurs independent of IL-1beta and IL-18 and is targetable by BET inhibitors. *Science Advances, 11*(28), eadv0079. https://doi.org/10.1126/sciadv.adv0079.

Simpson, D. S., Pang, J., Weir, A., Kong, I. Y., Fritsch, M., Rashidi, M., ... Vince, J. E. (2022). Interferon-γ primes macrophages for pathogen ligand-induced killing via a caspase-8 and mitochondrial cell death pathway. *Immunity, 55*(3), 423–441.e429. https://doi.org/10.1016/j.immuni.2022.01.003.

Skorda, A., Lauridsen, A. R., Huhtinen, K., Lahtinen, A., Senkowski, W., Oikkonen, J., ... Kallunki, T. (2023). Quantification of cell death and proliferation of patient-derived ovarian cancer organoids through 3D imaging and image analysis. *STAR Protocols, 4*(4), 102683. https://doi.org/10.1016/j.xpro.2023.102683.

Takasato, M., Er, P. X., Chiu, H. S., Maier, B., Baillie, G. J., Ferguson, C., ... Little, M. H. (2015). Kidney organoids from human iPS cells contain multiple lineages and model human nephrogenesis. *Nature, 526*(7574), 564–568. https://doi.org/10.1038/nature15695.

Takashima, S., Martin, M. L., Jansen, S. A., Fu, Y., Bos, J., Chandra, D., ... Hanash, A. M. (2019). T cell-derived interferon-γ programs stem cell death in immune-mediated intestinal damage. *Science Immunology, 4*(42), https://doi.org/10.1126/sciimmunol.aay8556.

Tan, T., Hirokawa, Y., Clarke, J., Sakthianandeswaren, A., & Sieber, O. M. (2022). Low-viscosity matrix suspension culture for human colorectal epithelial organoids and tumoroids. *Bio-protocol, 12*(8), e4394. https://doi.org/10.21769/BioProtoc.4394.

Tan, T., Mouradov, D., Gibbs, P., & Sieber, O. M. (2024). Protocol for generation of and high-throughput drug testing with patient-derived colorectal cancer organoids. *STAR Protocols, 5*(2), 103090. https://doi.org/10.1016/j.xpro.2024.103090.

Tan, T., Mouradov, D., Lee, M., Gard, G., Hirokawa, Y., Li, S., ... Sieber, O. M. (2023). Unified framework for patient-derived, tumor-organoid-based predictive testing of standard-of-care therapies in metastatic colorectal cancer. *Cell Reports Medicine, 4*(12), 101335. https://doi.org/10.1016/j.xcrm.2023.101335.

Tong, L., Cui, W., Zhang, B., Fonseca, P., Zhao, Q., Zhang, P., ... Lundqvist, A. (2024). Patient-derived organoids in precision cancer medicine. *Med, 5*(11), 1351–1377. https://doi.org/10.1016/j.medj.2024.08.010.

Wallach, D., Kang, T. B., & Kovalenko, A. (2014). Concepts of tissue injury and cell death in inflammation: A historical perspective. *Nature Reviews. Immunology, 14*(1), 51–59. https://doi.org/10.1038/nri3561.

Wang, R., Li, H., Wu, J., Cai, Z.-Y., Li, B., Ni, H., ... Mo, W. (2020). Gut stem cell necroptosis by genome instability triggers bowel inflammation. *Nature, 580*(7803), 386–390. https://doi.org/10.1038/s41586-020-2127-x.

Wang, Z., Zhao, S., Lin, X., Chen, G., Kang, J., Ma, Z., ... Xiang, D. (2022). Application of organoids in carcinogenesis modeling and tumor vaccination [review]. *Frontiers in Oncology, 12, 2022.* https://doi.org/10.3389/fonc.2022.855996.

Watanabe, S., Kobayashi, S., Ogasawara, N., Okamoto, R., Nakamura, T., Watanabe, M., ... Yui, S. (2022). Transplantation of intestinal organoids into a mouse model of colitis. *Nature Protocols, 17*(3), 649–671. https://doi.org/10.1038/s41596-021-00658-3.

Yin, X., Farin, H. F., van Es, J. H., Clevers, H., Langer, R., & Karp, J. M. (2014). Niche-independent high-purity cultures of Lgr5+ intestinal stem cells and their progeny. *Nature Methods, 11*(1), 106–112. https://doi.org/10.1038/nmeth.2737.

CHAPTER NINE

Short-term culture of human endoscopic specimens for the study of chronic inflammation

Julia Werner[a,b], Laura Hennig[c], Laura Schiller[a,b], Albrecht Hoffmeister[d], Sebastian Prill[d], Armin Frille[e], Hubert Wirtz[e], Ngoc Anh Hoang[f,g], Florian Lordick[g], Cica Vissiennon[a,b], and Sonja Kallendrusch[f,*]

[a]Medical Faculty, Institute of Medical Physics and Biophysics, Leipzig University, Leipzig, Germany
[b]Repha GmbH Biologische Arzneimittel, Langenhagen, Germany
[c]Institute of Anatomy, University of Leipzig, Leipzig, Germany
[d]Division of Gastroenterology, Department of Medicine II, University of Leipzig Medical Center, Leipzig, Germany
[e]Division of Respiratory Medicine, Department of Medicine II, University of Leipzig Medical Center, Leipzig, Germany
[f]Institute of Clinical Research and Systems Medicine, HMU Health and Medical University Potsdam, Potsdam, Germany
[g]Division of Oncology, Department of Medicine II, University of Leipzig Medical Center, University Cancer Center Leipzig (UCCL), Leipzig, Germany
*Corresponding author. e-mail address: sonja.kallendrusch@hmu-potsdam.de

Contents

Methods in Cell Biology, Volume 210
ISSN 0091-679X, https://doi.org/10.1016/bs.mcb.2026.05.014

Abstract

Inflammation and cell death are central to both chronic inflammatory disorders and cancer progression, influencing tissue homeostasis, therapeutic response, and resistance. Effective preclinical models that preserve the human microenvironment are required to study these processes, guide personalized therapy or to test pharmacologically active substances. We developed *patient-derived tissue slice cultures* from endoscopic specimens (ePDTC) such as gastrointestinal and pulmonary cancers, as well as inflamed mucosa from patients with Inflammatory bowel disease (IBD). These tissue cultures preserve epithelial, stromal, and immune compartments, enabling analysis of inflammatory signalling and cell death pathways while allowing parallel pharmacological testing. PDTCs maintained tissue cytoarchitecture and cell-cell interactions, permitting within-patient drug response profiling and capturing inter- and intra-patient heterogeneity. This enabled mechanistic investigation of inflammation-driven resistance and individualized assessment of chemotherapies, immunotherapies, and anti-inflammatory agents. Here, we describe a protocol for establishing ePDTC from diverse anatomical sites to provide a translational platform to study immune regulation and therapy response in human cancers and inflammatory diseases, supporting development of personalized treatment strategies.

1. Introduction

Cancer and chronic inflammatory disorders share a central feature: dysregulated tissue homeostasis driven by inflammatory signaling and cell death (Kayagaki et al., 2024). In malignancies such as gastric, lung, and colorectal cancer, inflammation and immune cell death are not only hallmarks of tumor progression but also key determinants of therapy resistance (Ma et al., 2022; Spella & Stathopoulos, 2021; Sui et al., 2022). Similarly, in chronic inflammatory diseases like inflammatory bowel disease (IBD), persistent cycles of damage and repair sustain disease activity and complicate therapeutic interventions (Stallmach et al., 2025). Despite advances in systemic therapies, including chemotherapy, immunotherapy, and targeted drugs, broad anti-inflammatory or cytotoxic approaches have shown relatively limited efficacy, high relapse rates, and substantial toxicity. Thus underscoring the need for models that capture the complex interplay between immune regulation, cell death, and tissue repair.

Conventional cell lines and co-culture systems provide mechanistic insights but fail to reflect the structural and immunological complexity of human tissues. Organoids and spheroids advance toward three-dimensional modelling but remain limited in immune and stromal representation (Dao et al., 2022; Adelfio et al., 2025; Yip et al., 2023; Goldrick et al., 2023). By contrast, patient-derived tissue slice cultures (PDTCs) obtained from endoscopic or bronchoscopic biopsies preserve native cytoarchitecture, inflammatory networks, and tumor – immune interactions (Molleken et al., 2025; Sonnichsen et al., 2018; Hennig et al., 2025; Vaira et al., 2010; Majumder et al., 2015; Prill et al., 2019; Junk et al., 2021; Husstegge et al., 2021; Bruckner et al., 2021). This allows functional assessment of tissue responses, including regulated cell death pathways (apoptosis, necroptosis, pyroptosis, ferroptosis), within a clinically relevant microenvironment (Molleken et al., 2025; Hennig et al., 2025; Majumder et al., 2015; Prill et al., 2019; Junk et al., 2021; Husstegge et al., 2021; Bruckner et al., 2021).

Endoscopic and bronchoscopy sampling provides minimally invasive access to viable tissue from gastrointestinal and pulmonary cancers, as well as inflamed mucosa in IBD or chronic obstructive pulmonary disease (COPD). These samples can be processed into precision-cut tissue slices, enabling parallel treatment comparisons both within individual biopsies (multiple slices per specimen) and across biopsies from the same patient, thereby controlling for interindividual heterogeneity, tissue resistance and individual susceptibility. Multiple analytical endpoints can be applied to the same slice, maximizing information yield from limited biopsy material. Importantly, PDTCs retain epithelial, stromal, and immune compartments, allowing investigation of how microenvironmental factors influence inflammatory signalling, barrier integrity, immune cell death, and drug responsiveness (Hennig et al., 2025) (Fig. 1).

Applying this approach across multiple cancer entities and inflammatory conditions enables to study disease-specific inflammatory and cell death processes in human tissues, and functional testing of pharmacological interventions, including chemotherapy, immunotherapy, and experimental compounds. This dual perspective is essential to bridge the gap between preclinical models and individualized therapy, advancing precision medicine in both oncology and inflammatory disease.

Here, we describe the culture of biopsy-derived tissues from inflammatory bowel disease and cancers, including colon, gastric, and lung cancer. In addition to histological assessment, supernatant collection and application for multiplex ELISA are described.

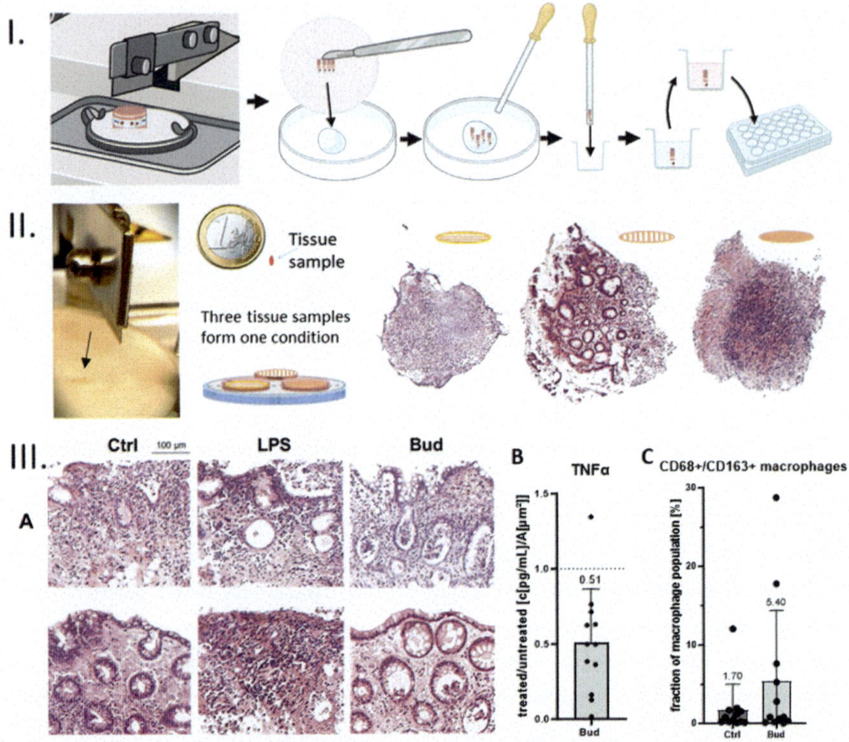

Fig. 1 I. Tissue processing and analysis I. Schematic tissue processing. Placement of the biopsy on the cutting stage of the tissue chopper (350 µm setting). After sectioning, carefully separate the slices from the filter paper into preparation media, and transfer slices onto membrane inserts for air–liquid interface culture. Created by Biorender. **II. Tissue variability and alignment.** Real-size (smallest sample demonstrated) comparison illustrates the possible alignment of individual slices. Due to the small tissue amount careful selection and distribution across experimental conditions is required. Shown is the high heterogeneity of specimens obtained from a single gastric cancer patient. **III. Experimental stimulation and analysis.** (A) Stimulation of two patient-derived specimens with IBD (upper row: active inflammation; lower row: low inflammatory activity). (B) Measurement of supernatants: TNF-α levels decrease after 24 h of treatment with 1 µM budesonide (Bud). Data are shown as fractional change (treated/untreated) in released concentration per area [pg/mL × µm^2], normalized to untreated samples (set at x = 1); mean ± SD, n = 12. (C) Histological analysis: Immunofluorescence staining of CD68$^+$/CD163$^+$ macrophages after 24 h without treatment (Ctrl) or with 1 µM budesonide (Bud). *Data shown as fraction of the total macrophage population; mean ± SD, n = 12.*

2. Materials

Ethics

Handling human tissue needs an approval by the local ethics committee and proper patient pseudonymization protocols must be followed throughout the study. This method only uses tissue that is not required for routine diagnosis. Prior every experiment, it is necessary to verify that the patient provided informed consent and that this consent is properly documented.

General disposables

- Petri dishes 100 mm (#664160, Greiner)
- Petri dishes 60 mm (#93060, TPP)
- Pasteur pipette 150 mm (#7691060, Labsolute) (Note 1)
- Culture plates, 24 (# 92006, TPP)
- Millicell® Inserts 12 mm (#PICM01250, Merck Millipore)
- filter paper (#205005, Macherey Nagel Ø 55 mm)
- razor blade, microtome blade
- scalpel blade (B.Braun No. 11 and No. 15)
- Rubber Bulbs for Pasteur pipette (silicone, autoclavable Carl Roth GmbH+Co. KG, #C530.1)
- glass slides (Thermo Fischer Scientific, SuperFrost Slides)

Instruments:

- 12–15 cm forceps (rounded tip, straight or curved)
- 12–15 cm forceps (pointed tip, curved)
- Blade holder (blades no.3 or no.4)

Reagents:

- Advanced D-MEM /F12 (#12634010, Gibco)
- Fetal bovine serum (#10270, Gibco)
- L-Glutamin (#25030-081, Gibco)
- (Phenol-free) RPMI 1640 (Gibco™, #32404014)
- DPBS (#14190094, Gibco)
- PFA 16 % (#28908, Thermo Fischer)
- Budesonide (#B7022, Sigma-Aldrich)
- Paraformaldehyde (16 % methanol-free PFA 10 mL glass ampules, Thermo Fischer, #28908)
- Paraffin wax melting point 56 °C (Carl Roth GmbH + Co.KG, ROTI® Plast #6642)

Optional:
- Penicillin/Streptomycin (#15140-122, Gibco)
- Amphotericin B (#0246.1, Carl Roth)
- Eosin Y (#7089.1, Carl Roth)

2.1 Equipment
- Tissue chopper (Cavey Laborator Engineering Co. Ltd., McIlwain Tissue Chopper, Model TC752)
- Stereomicroscope (EVIDENT Europe GmbH (former Olympus), Model SZX16)
- Biological safety cabinet (Fisher Scientific GmbH, Model Herasafe 2030i)
- Laminar flow bench (Nordic Labtech AB, Model nSAFE Horizontal)
- Microwave or Pressure Cooker (Aptum Biologics Ltd., Model 2100 Antigen Retriever)
- Vacuum Tissue Processor (Leica Biosystems, Model Tissue Processor Tp1020 Histokinette)
- Paraffine Embedding Station (Leica Biosystems, Model HistoCore Arcadia H)
- Microtome (Leica Biosystems, Model SM2010 R Sliding Microtome)
- Fluorescence microscope (Olympus BX51)
- Luminescence reader (Luminex® FLEXMAP 3D®)

2.2 Software
- ImageJ (http://imagej.hih.gov/ij)
- Qupath (https://qupath.github.io)
- 3DHISTECH Pannoramic Viewer (if using digital scanner)
- GraphPad Prism or equivalent statistical software

3. Methods

3.1 Patient samples and transport

Ethical requirements

1. Obtain institutional review board approval: Obtain ethical approval for human tissue use according to national and Declaration of Helsinki guidelines
2. Ensure written informed consent is obtained from all patients prior to sample collection and implement proper patient anonymization and tissue tracking protocols throughout the study.

Media recipes

- Preparation media
  - RPMI 1640
  - 1 % Penicillin/streptomycin
  - 1 % Amphotericin B
- Culture media
  - Advanced D-MEM/F12
  - 15 % FBS
  - 1 % Penicillin/streptomycin
  - 1 % Amphotericin B (Note 2)
  - 1 % L-Glutamin

Tissue acquisition

- Immediately place specimens in sterile 15 mL conical tubes containing 3–5 mL cold preparation medium.
- Transport samples on ice to the laboratory within 2–4 h after collection.
- Process samples immediately upon arrival. (Note 3)

3.2 Tissue processing

The tissue must be processed quickly up to the point of culture in order to maintain tissue viability.

1. Pre-warm all media to 37 °C and prepare sterile workspace under laminar flow hood.
2. Cover the tissue chopper platform with a stack of filter paper and put a sterile filter paper into a clean petri dish lid (60 mm).
3. Fasten the razor blade onto the tissue chopper.
4. Adjust/ check settings to 350 μm slicing steps and cutting speed. (Note 4)
5. Remove biopsies from the transportation tube (15 mL tubes) and transfer into clean petri dish (60 or 100 mm). (Note 5)
6. Carefully wash the biopsy specimen 2 times in preparation media. Separate individual biopsies using forceps.
7. Transfer one single biopsy to the centre of the prepared filter paper; orient tissue appropriately:
   a. For intestinal tissue: mucosal surface facing upward.
   b. For cancer tissue: all organotypic structures should be cut by the blade.
8. Place the filter paper with the tissue onto the tissue chopper stage and wet the tissue immediately with preparation medium to prevent desiccation.

9. Re-position the cutting table as needed.
10. Stabilize the filter paper with forceps during the cutting process.
11. Start tissue chopper and avoid blunt impacts during the cutting process.
12. Transfer the filter paper with the tissue into a new petri dish filled with preparation media.
13. Turn the filter paper upside down just above the medium- the tissue should drop into the medium. If necessary, help with a scalpel by carefully removing the tissue from the paper.
14. Transfer each sliced biopsy to a new 60 mm petri dish (label the lid).
15. Stereomicroscope, carefully separate individual slices using fine forceps and scalpel. (Note 6)
16. Control each slice for the thickness under the Stereomicroscope.

3.3 Culture conditions

Culture.

- Add 250 μL culture medium into a 24 well culture plate. (Note 7)
- Place each slice onto a membrane insert using the blunt Pasteur pipette; ensure the slices are not in contact with the edges of the insert. (Note 8)
- Remove excess medium from slice surface using a pointed Pasteur pipette.
- Avoid air bubble formation by gently moving inserts with sterile forceps.
- Transfer the well-plates with the tissue cultures to the incubator (36 °C, 5 % CO_2, standard humidity). (Note 9)
- Replace medium after 2−4 h to remove excessive accumulation of inflammatory molecules due to culture preparation.
- Start pharmacological intervention directly or after 24 h, depending on the experimental set-up.

Baseline sample collection (optional but recommended).

- Set aside one slice from each biopsy for immediate fixation as baseline control (t = 0 h) (Note 10)
- Fixate selected slice immediately in 1 mL 4 % paraformaldehyde in PBS for 16−24 h at 4 °C and continue as described below

3.4 Treatment protocols

Treatment preparation

- Prepare treatment compounds in culture medium at desired concentrations.
- Include appropriate vehicle controls (DMSO/Ethanol, NaCl etc.) for each experiment.

Treatment application

- Remove culture plates from incubator and work under sterile conditions.
- Lift the insert with sterile forceps and remove medium without disturbing tissue slices.
- Add 250 μL fresh culture medium with or without treatment compounds.
- Repeat this for each single well separately to avoid desiccation.
- Return plates to the incubator immediately after medium exchange.
- Initiate treatment period of 24−48 h as per experimental design.

3.5 Sample collection and analysis

Supernatant collection.

- At designated time points, carefully collect culture medium from lower chamber using a sterile pipette.
- Transfer supernatant to sterile 500 μL microcentrifuge tubes.
- Centrifuge samples at 600 g for 5 min at room temperature to remove cellular debris.
- Transfer cleared supernatant to fresh tubes and store at −80 °C until analysis.
- Avoid freeze-thaw cycles; aliquot samples if multiple analyses are planned.

Cytokine analysis

- Thaw supernatant samples on ice and vortex gently before analysis.
- Perform multiplex cytokine analysis using customized Luminex assays (R&D Systems) according to manufacturer protocols.
- Analyze immune-associated mediators including TNF-α, IFN-γ, IL-1β, IL-2, IL-10 and additional inflammatory markers. (Note 11)
- Read plates using Luminex® FLEXMAP 3D® system with appropriate standard curves and quality controls.

Tissue collection for histology.

- Fixate tissue in 1 mL 4 % paraformaldehyde (PFA) in PBS for 16−24 h at 4 °C. Carefully add PFA on top of the insert.
- Wash twice in DPBS for 1 h
- Optional: Pre-stain tissue to improve recovery after embedding (Note 12)
  - Dehydrate in EtOH 70 % for 30 min
  - Carefully transfer the slices from the inserts to the bottom of the respective wells using a brush.
  - Stain with Eosin Y (1 % in EtOH 96 %) for 30 min
  - Wash with EtOH 100 % for 10 min
- Process for paraffin embedding within 7 days of collection.

Histological staining (Note 13).

- Prepare 5 μm paraffin sections using standard microtome techniques.
- Perform immunohistochemistry (IHC) using Diaminobenzidine (DAB) detection or immunofluorescence for target proteins. Block with 5 % normal goat serum + 1 % BSA in PBS/Triton 0.5 h (paraffin sections or 4 h whole mount).
- Primary antibodies (overnight, 4 °C, PBS/Triton)
- Secondary antibodies (2 h, room temperature on shaker, PBS/Triton)
- Nuclear counterstain with DAPI.
- Mount and image within 7days.

Data analysis

- Data preprocessing: Cytokine concentrations were normalized to the corresponding tissue slice area measured on histological sections to account for tissue size variations, yielding concentrations in pg/mL per μm^2. Data underwent logarithmic transformation to achieve normal distribution, verified through Shapiro-Wilk testing. (Note 14)
- Statistical modeling: Linear mixed–effects models (LMM) were calculated to account for the nested data structure, with patient included as random intercept and treatment (e.g., budesonide) as fixed effect. Degrees of freedom and p-values for fixed effects were estimated by Satterthwaite's method.

4. Notes

1. Creating blunt Pasteur pipettes:
   a. Use diamond engraving pen to cut around the base of the Pasteur pipette tip.
   b. Carefully break the pasteur pipette tip off.
   c. Cover the sharp-edged side with the rubber bulbus while tissue processing.
2. Proceed with transferring the tissue punches onto the insert's PTFE-membrane.
3. Addition of Amphotericin B has impact on macrophage activation. Intensive washing of the native specimen and of the tissue slices reduces contamination events.
4. Inflammatory tissue should be processed as soon as possible not exceeding 2 h. Tumor specimen can remain longer in the preparation buffer; however, transportation tubes must only be filled with 2–3 mL preparation media.

5. Speed and force should be adjusted to the tissue and requires some exercise. Colon biopsy specimen can curl during the cutting process, ensure proper hydration and check blade sharpness. Gentle pressure with forceps may help to maintain flat positioning. Do not quench or push the tissue.
6. Pour the medium out into a petri dish, if the tissue sticks to the wall, use sterile and blunt forceps to "carry" the tissue. Do not quench, pierce or push the tissue.
7. Tissue orientation is important to consider. The tissue, although small, needs to be oriented properly, so that all organotypic structures of interest are represented in one slice.
8. Although the tissue specimen are small, thickness of the slices needs to be controlled. Here we describe the method to cultivate and measure single tissue slice reactivity. Using tumor specimen, several endoscopic specimen slices can be cultures on 6-well inserts, creating replicates of one condition and considering tissue heterogeneity.
9. Placing tissue punches onto insert area
   a. Pick up each tissue punch with preparation media using a blunt Pasteur pipette
   b. Let the tissue punch settle within the liquid
   c. Carefully squeeze the rubber bulb to create convex liquid surface
   d. Carefully place the tissue punch onto PTFE membrane insert
      i. Avoid insert edge by 1–2 mm, media can pool and lift tissue
      ii. Leave at least 2–5 mm space around each tissue punch
      iii. Maximum of 6 × 3 mm tissue punches per insert

4.1 Remove excess preparation media

1. Tissue can be cultivated at 37 °C, however to avoid immunological stimulation temperature of the tissue incubator should be set at 36 °C.
2. The native tissue is an important reference tool to determine culture artefacts and should not be taken from the tissue selected for disposal. Experimental outcome can be correlated to the native tissue, directly referring to clinical indications.
3. We analyzed 24 immune-associated mediators including TNF-α, IL-1β, IFN-γ, IL-2, IL-10 as well as Angiogenin, MIF, CXCL5/ENA-78, VEGF, MCP-1, CXCL9/MIG, G-CSF, GDF-15, GM-CSF, Lipocalin-2, TFF3, uPAR, IL-4, IL-17/IL-17A, IL-23, IL-33, IL-6, IL-8, and Serpin-E1.
4. Embedding of small tissues needs to be handled with care. Therefore, we suggest staining of whole mounts with eosin to find tissue specimen after

processing in the embedding machine. To avoid loss of tissue slices manual embedding is recommended. Due to the small size of the tissue slices embedding each slice in a small volume prior embedding into the cutting block can help to quickly find the plane and tissue, beside eosin staining. The tissue samples on the glass slides should also be marked by a glass marker next to the tissue to find the specimen quickly under the microscope.

5. The tissue section area appeared to vary considerably (e.g., sections from the edges of a biopsy were smaller than the center sections), which is why the released mediator concentration was normalized to this area. With regard to sample thickness, it is assumed that all samples have the set thickness of 350 μm, and therefore this value was not additionally included in the normalization.
6. The medium volume in 6-well plates requires at least 900 μl, therefore supernatant might yield lower cytokine secretion and a sampling time of 48 h is recommended. However, secretion marker should be adjusted to the amount of tissue cultured. For mediator release investigations we recommend to use 24 wells and to culture each slice individually.

5. Concluding remarks

The patient-derived tissue culture (PDTC) platform offers a translational system with minimal patient burden, as specimens can be obtained during routine endoscopic or bronchoscopic procedures. By preserving epithelial, stromal, and immune compartments, ePDTCs retain the structural and cellular complexity that is essential for dissecting disease mechanisms and capturing patient-specific heterogeneity (Husstegge et al., 2021). Preservation of immunocompetence in paraffin-embedded ePDTCs allow for the analysis of immune cell infiltration, epithelial disruption and inflammation profiling, which is essential for research on chronic inflammatory diseases such as IBD (Mobbs et al., 2024). Importantly, they permit parallel pharmacological testing, enabling individualized assessment of therapeutic responses under conditions that closely reflect the human microenvironment. Further culture systems like, organoid cultures, though expandable for weeks, lack resident immune cells, organotypic tissue architecture and stromal interactions limiting their utility for studying immune-epithelial interactions (Parente et al., 2024). Another approach, using whole biopsy explants, yield insufficient tissue for within-sample comparative testing, precluding detailed pharmacological assessment (Vadstrup et al., 2016).

Still, in contrast to PDTCs from surgical resections, maintained for up to 6 days in culture (Molleken et al., 2025; Majumder et al., 2015; Prill et al., 2019; Junk et al., 2021; Husstegge et al., 2021; Bruckner et al., 2021), ePDTCs are less stable under culture conditions and should only be investigated for 24h-48h (Hennig et al., 2025; de Graaf et al., 2010). This time window, although short, allow for a balance between experimental feasibility and tissue viability, supporting mechanistic insights into therapy modulation.

Nevertheless, the approach carries limitations, including the restricted culture duration, potential artifacts introduced during tissue handling, and the requirement for immediate processing after biopsy collection. Some points might be addressed by perfusion systems and integration of blood derived lymphocytes and signaling molecules as well as extracellular vesicles. Despite these constraints, ePDTCs represent a powerful ex vivo platform to advance personalized therapy development and to bridge the gap between preclinical models and clinical translation (Hennig et al., 2025).

Acknowledgements

The authors thank the assistance of the University of Leipzig Medical Center, Division of Gastroenterology, Department of Medicine II, Leipzig, Germany.

Competing interests

JW, LS and CV were employed by Repha GmbH Biologische Arzneimittel. FL reports research grants for his institution from: Astra Zeneca, Beigene, BMS, Daiichi Sankyo and Gilead, and personal fees for educational and advisory activities from: Amgen, ArtTempi. Astellas, Astra Zeneca, Bayer, BMS, Boehringer Ingelheim, Iomedico, Daiichi Sankyo, Eli Lilly, Gilead, Elsevier, Incyte, Medscape, MedUpdate, Merck Serono, MSD, PAGE, Roche, Servier, StreamedUp, VJ Oncology, all outside the submitted work. AF reports grants from the Federal Ministry of Education and Research (BMBF), Germany (FKZ 01EO1501, IFB Adiposity Diseases), the "Mitteldeutsche Gesellschaft für Pneumologie (MDGP) e.V." (2018-MDGP-PA-002), the Medical Faculty, University of Leipzig (934100–012), the "Novartis Foundation", and the ERA PerMed (ERAPerMed_324; Horizon 2020 Research and Innovation Framework Programme).

References

Adelfio, M., et al. (2025). Engineered tissue models to decode host-microbiota interactions. *Advanced Science (Weinh), 12*(23), e2417687.

Bruckner, L., et al. (2021). FOXM1 inhibition in ovarian cancer tissue cultures affects individual treatment susceptibility ex vivo. *Cancers (Basel), 13*(5).

Dao, V., et al. (2022). Immune organoids: From tumor modeling to precision oncology. *Trends Cancer, 8*(10), 870–880.

de Graaf, I. A., et al. (2010). Preparation and incubation of precision-cut liver and intestinal slices for application in drug metabolism and toxicity studies. *Nature Protocols, 5*(9), 1540–1551.

Goldrick, C., et al. (2023). 3D multicellular systems in disease modelling: From organoids to organ-on-chip. *Frontiers in Cell and Developmental Biology, 11*, 1083175.

Hennig, L., et al. (2025). Personalized drug stratification using endoscopic samples to assess ex vivo gastric cancer tissue susceptibility to chemotherapy and immune checkpoint inhibitors. *Clinical and Experimental Medicine, 25*(1), 188.

Husstegge, M., et al. (2021). PD-1 inhibition in patient derived tissue cultures of human gastric and gastroesophageal adenocarcinoma. *Oncoimmunology, 10*(1), 1960729.

Junk, D., et al. (2021). Human tissue cultures of lung cancer predict patient susceptibility to immune-checkpoint inhibition. *Cell Death Discovery, 7*(1), 264.

Kayagaki, N., Webster, J. D., & Newton, K. (2024). Control of cell death in health and disease. *Annual Review of Pathology, 19*, 157–180.

Ma, E. S., et al. (2022). Immune evasion mechanisms and therapeutic strategies in gastric cancer. *World Journal of Gastrointestinal Oncology, 14*(1), 216–229.

Majumder, B., et al. (2015). Predicting clinical response to anticancer drugs using an ex vivo platform that captures tumour heterogeneity. *Nature Communications, 6*, 6169.

Mobbs, C. L., Darling, N. J., & Przyborski, S. (2024). An in vitro model to study immune activation, epithelial disruption and stromal remodelling in inflammatory bowel disease and fistulising Crohn's disease. *Frontiers in Immunology, 15*, 1357690.

Molleken, J., et al. (2025). Artemisinin derivatives differently affect cell death of lung cancer subtypes by regulating GPX4 in patient-derived tissue cultures. *Cell Death Discovery, 11*(1), 256.

Parente, I. A., Chiara, L., & Bertoni, S. (2024). Exploring the potential of human intestinal organoids: Applications, challenges, and future directions. *Life Sciences, 352*, 122875.

Prill, S., et al. (2019). Tumor-associated macrophages and individual chemo-susceptibility are influenced by iron chelation in human slice cultures of gastric cancer. *Oncotarget, 10*(46), 4731–4742.

Sonnichsen, R., et al. (2018). Individual susceptibility analysis using patient-derived slice cultures of colorectal carcinoma. *Clinical Colorectal Cancer, 17*(2), e189–e199.

Spella, M., & Stathopoulos, G. T. (2021). Immune resistance in lung adenocarcinoma. *Cancers (Basel), 13*(3).

Stallmach, A., et al. (2025). Treatment of severe flares in Crohn's disease and ulcerative colitis. *Innere Medizin (Heidelberg, Germany), 66*(1), 22–30.

Sui, Q., et al. (2022). Inflammation promotes resistance to immune checkpoint inhibitors in high microsatellite instability colorectal cancer. *Nature Communications, 13*(1), 7316.

Vadstrup, K., et al. (2016). Validation and optimization of an ex vivo assay of intestinal mucosal biopsies in Crohn's disease: Reflects inflammation and drug effects. *PLoS One, 11*(5), e0155335.

Vaira, V., et al. (2010). Preclinical model of organotypic culture for pharmacodynamic profiling of human tumors. *Proceedings of the National Academy of Sciences of the United States of America, 107*(18), 8352–8356.

Yip, S., Wang, N., & Sugimura, R. (2023). Give them vasculature and immune cells: How to fill the gap of organoids. *Cells Tissues Organs, 212*(5), 369–382.

CHAPTER TEN

In vitro protocols for inducing and assessing endotoxin tolerance in human and murine macrophages

Aoife P. Smith, Mathew J. Choaji, and Emma M. Creagh*

School of Biochemistry and Immunology, Trinity Biomedical Sciences Institute, Trinity College Dublin, Dublin, Ireland

*Corresponding author. e-mail address: ecreagh@tcd.ie

Contents

Abstract

Macrophages are key innate immune effector cells capable of responding to microbial factors such as lipopolysaccharide (LPS) through activation of inflammatory pathways. However, repeated LPS exposure can lead to endotoxin tolerance, a reprogrammed state where macrophages suppress pro-inflammatory cytokine production while preserving antimicrobial functions. The reduction in inflammatory responses during endotoxin tolerance serves to preserve tissue integrity and macrophage viability, although it can also result in immunosuppression, as observed in sepsis. An enhanced understanding of the molecular mechanisms which govern endotoxin tolerance in macrophages could unlock new ways to modulate innate immunity. Here we provide detailed protocols for inducing and measuring tolerance

Methods in Cell Biology, Volume 210
ISSN 0091-679X, https://doi.org/10.1016/bs.mcb.2026.05.013

in murine bone marrow-derived macrophages (BMDMs) and human PMA-differentiated THP-1 macrophages, using ELISA-based quantification of IL-1β and IL-6 as functional readouts. These methods provide robust in vitro systems to study endotoxin tolerance.

1. Introduction

Macrophages are critical effector cells of the innate immune system. Serving at the frontline of immune defence, macrophages are continuously exposed to microbial components and other damage or pathogen associated molecular patterns (DAMPS/PAMPs). PAMPs are conserved microbial structures that serve as danger signals to the immune system. They are recognized by pattern recognition receptors (PRRs), which activate innate immune responses to detect and eliminate invading pathogens. Among the most potent microbial PAMP stimuli is lipopolysaccharide (LPS), a major component of the outer membrane of Gram-negative bacteria. Extracellular detection of LPS by macrophages occurs via the PRR, Toll-like receptor 4 (TLR4), which triggers inflammatory signalling via the transcription factor NFκB (Hoshino et al., 1999; Luo et al., 2025; Rhee & Hwang, 2000). Macrophage detection of intracellular LPS occurs via activation of inflammatory caspases (caspase-11 in mice, caspases-4 and -5 in humans) which triggers pyroptosis, a potently inflammatory form of cell death (Agnew et al., 2021; Smith & Creagh, 2022). Therefore, macrophage encounters with LPS provoke strong inflammatory responses characterized by cytokine release and, in some contexts, inflammasome activation and pyroptotic cell death (Kawai et al., 2024). In tissues such as the gut, liver, and lung, repeated exposure of tissue resident macrophages to LPS is common, raising the question of how these cells avoid self-destruction as well as tissue-destruction, while preserving antimicrobial competence. The phenomenon of LPS tolerance provides an answer to this question.

LPS tolerance describes the state in which macrophages, after an initial repeated or prolonged exposure to endotoxin (LPS), respond less vigorously to subsequent stimulation (Roszak et al., 2025; Seeley & Ghosh, 2017). This is not simply a failure of the cell to respond, but rather a controlled reprogramming of the response profile (Gillen et al., 2021; Mages et al., 2007). LPS tolerant macrophages suppress expression of genes encoding pro-inflammatory cytokines such as IL-1β and IL-6, while maintaining or even enhancing expression of antimicrobial effectors and regulatory mediators (Bambouskova et al., 2021; Gillen et al., 2021; Li et al., 2023). This selective dampening ensures that repeated activation does not trigger excessive inflammation that could compromise tissue integrity or macrophage viability (Widdrington et al., 2018).

From the perspective of cell survival, tolerance functions as a protective adaptation. Strong LPS responses can prime macrophages for pyroptosis, apoptosis, or necroptosis, all of which, while beneficial in clearing infection under acute conditions, would be maladaptive if triggered repeatedly in tissue-resident populations (Seeley & Ghosh, 2017). By curbing pro-death pathways downstream of persistent TLR4 stimulation, tolerance preserves macrophage survival during recurrent or chronic exposure (Widdrington et al., 2018).

Clinically, the balance between protective tolerance and harmful immunosuppression is critical. In sepsis, systemic LPS tolerance is often interpreted as a cause of "immunoparalysis," but at the cellular level, it likely evolved to prevent macrophage death and maintain phagocyte populations during repeated endotoxin exposure. At mucosal barriers, where Gram-negative bacteria are abundant, tolerance may be essential for long-term macrophage persistence and tissue homeostasis (Roszak et al., 2025).

The importance of studying LPS tolerance lies in its dual role as both a protective and potentially pathological mechanism. On the one hand, it preserves macrophage populations and tissue homeostasis in environments with frequent microbial exposure, such as the gut. On the other, when tolerance is systemic or prolonged, as in sepsis or chronic infection, it can contribute to immune suppression and poor pathogen clearance (Chen et al., 2023). Understanding how tolerance prevents macrophage death while avoiding detrimental immunosuppression may reveal therapeutic opportunities to fine-tune innate responses. To achieve this, developing easy and reproducible methods to model and measure tolerance in vitro is crucial.

Here we describe such an approach involving in vitro macrophage stimulation protocols in both human and murine systems, combined with ELISA-based monitoring of pro-inflammatory cytokines IL-6 and IL-1β as readouts. This straightforward system provides an experimental model of tolerance induction and can be extended by analysing specific genes and proteins in knockout or transgenic macrophages, thereby linking tolerance phenotypes to defined molecular pathways.

2. Materials (see Note 1)

2.1 Common disposables

- 0.22 μm sterile syringe filter (VWR #514-0061)
- 0.45 μm cell strainer (Greiner Bio-one CELLSTAR® #22363548)
- 15 cm^2 cell culture dish (Greiner Bio-one CELLSTAR® #639160)
- 20 mL sterile individually wrapped syringes (Terumo #ss+20L1)
- 25-gauge sterile individually wrapped needles (Terumo #2525R1)
- 5 mL, 10 mL, 25 mL serological pipettes (Greiner Bio-one CELLSTAR® #P7615, P7740, P7865)
- 50 mL conical tubes (Greiner Bio-one CELLSTAR® #227261)
- 6-, 12-, and 24-well cell culture plates (Greiner Bio-one CELLSTAR® #657160, 665180, 662160)
- 70% ethanol
- Autoclaved P1000, P200, P20, and P10 pipette tips (Corning #4868, #4845, #4080)
- Cell culture flasks, 75 cm^2 with filter screw cap (Greiner Bio-one CELLSTAR® #658175)
- Disposable scalpel – No. 23 (Swann-Morton #INS4753)
- MaxiSorp™ non-sterile high-binding 96-well plate (Thermo Fisher #456537)
- Polypropylene microcentrifuge tubes, 1.5 mL (Greiner Bio-one #616201)
- Sterile individually packed cell scraper (Greiner Bio-one #541070)

2.2 Cells and reagents

- 3, 3′, 5, 5′ tetramethyl benzidine (TMB) substrate set (Biolegend #421101)
- Assay Diluent: 1 % bovine serum albumin (BSA) in phosphate-buffered saline (PBS) (BSA Merck #A7030, PBS Gibco #14190)
- Bone Marrow cells (Murine) (*see* **Note 2**)
- Coating Buffer: $NaHCO_3$ 99.9917 mmol·L^{-1}, $NaCO_3$ 33.5849 mmol·L^{-1}, pH 9.5 (Merck # S5761, 223530)
- Complete DMEM: Dulbecco's Modified Eagle's Medium, high glucose (4.5 g·L^{-1}), GlutaMAX™ Supplement (Gibco™ #10566016), supplemented with 10 % FBS (Merck #F9665), 10,000 U·mL^{-1} penicillin, and 10 mg·mL^{-1} streptomycin (Gibco™ #15070063) (*see* **Note 3**)
- Complete RPMI 1640 medium: Gibco™ RPMI 1640 Medium, high glucose (4.5 g·L^{-1}), GlutaMAX™ Supplement (Gibco™ #61870036), supplemented with 10 % FBS (Merck #F9665), 10,000 U·mL^{-1} penicillin, and 10 mg·mL^{-1} streptomycin (Gibco™ #15070063) (*see* **Note 3**)
- ELISA MAX™ Deluxe Set Human IL-1β (Biolegend #437004)

- ELISA MAX™ Deluxe Set Mouse IL-6 (Biolegend #431304)
- ELISA MAX™ Standard Set Human IL-6 (Biolegend #430501)
- ELISA MAX™ Standard Set Mouse IL-1β (Biolegend #432601)
- Hybri-Max™ red blood cell (RBC) lysis buffer (Merck #R7757)
- L929 cell line (ATCC) (*see* **Notes 4)**
- L929 conditioned medium (CM) (*see* **Notes 5**)
- Lipopolysaccharides from Escherichia coli O55:B5 (Merck #L6529)
- Phorbol 12-myristate 13-acetate (PMA) (Merck #524400)
- Phosphate-buffered saline (PBS): NaCl 136.893 $mmol{\cdot}L^{-1}$, KCL 2.683 $mmol{\cdot}L^{-1}$, Na_2HPO_4 10.144 $mmol{\cdot}L^{-1}$, NaH_2PO_4 1.8 $mmol{\cdot}L^{-1}$ (Merck # 1613804, P9541, S3264, S3139)
- Stop solution: 2 $mol{\cdot}L^{-1}$ H_2SO_4 (Chem Lab # 11973303)
- THP-1 Monocyte cell line (ATCC)
- Trypan Blue (Merck #T8154)
- Virkon RelyON™ (Scientific Laboratory Supplies # CLE1552SP)

2.3 Equipment

- Centrifuge Legend T (Sorvall)
- Class II laminar flow hood Microflow Peroxide (Bioquell)
- Dissection Kit (Scientific Laboratory Supplies # 430-004)
- Haemocytometer (Hausser Scientific)
- Light Microscope (Leica)
- Microplate spectrophotometer SPECTRA max Pro (Molecular Devices)
- P1000, P200, P20, P10 micropipettes Pipeteman (Gilson)
- See-Saw Rocker (Thermo Fisher)
- Steri-cycle CO_2 incubator (Thermo Fisher)
- Water Bath (Clifton)

2.4 Software

- GraphPad Prism (v10.5.0)
- Microsoft Excel
- SoftMax Pro

3. Methods

3.1 Generation of bone marrow-derived macrophages (BMDM)

A timeline and summary of the BMDM differentiation protocol is presented in Fig. 1.

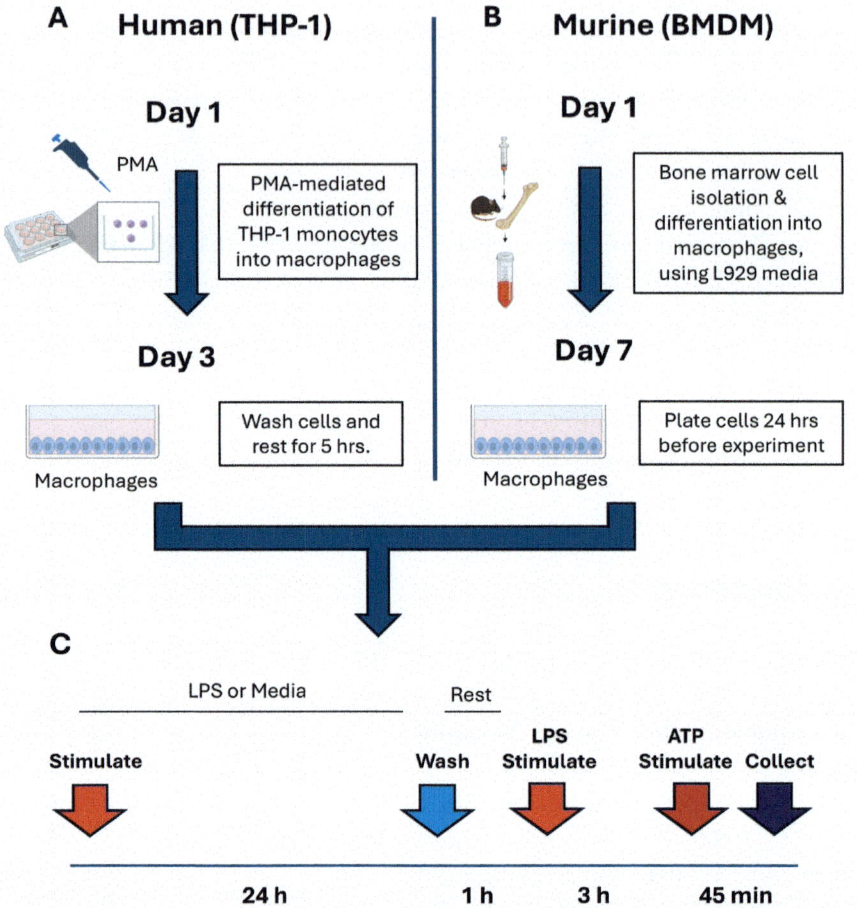

Fig. 1 Experimental Protocol and Timeline for the Induction of Tolerance in Human and Murine Macrophages. (A) Human macrophages are generated by differentiating THP-1 monocytes (plated at 1 × 106/mL) and differentiated with 1.6 mM PMA for 48 h. Following differentiation, cells are washed with PBS to remove residual PMA, replenished with fresh medium, and rested for 5 h prior to tolerance stimulation. (B) Murine macrophages are derived from bone marrow cells isolated from femur and tibia of mice. Cells are cultured for 6 days in M-CSF, provided by L929 cell-conditioned medium (20 %). Differentiated murine macrophages are plated at 1 × 106/mL and rested for 24 h prior to tolerance stimulation. (C) BMDM and PMA differentiated THP-1 cells are stimulated with LPS (24 h) to induce tolerised phenotype (T) or left untreated for non-tolerised phenotype (NT). Cells are washed with PBS and left to rest for 1 h before stimulating with LPS (100 ng/mL for BMDM, 10 ng/mL for THP-1) for 3 h and ATP (5 mM, 45 min) to activate the canonical inflammasome.

1. Euthanize mice humanely in accordance with institutional ethical guidelines (*see* **Note 6**).
2. Spray carcasses with 70 % ethanol. Make a small incision halfway up the back, gently lift the skin with a forceps and cut to remove the lower half of the skin and fur, exposing the lower limbs.
3. Using scissors, excise femurs and tibiae. Submerge the bones in complete DMEM (cDMEM) within a 50 mL conical tube (*see* **Note 7**).
4. Perform all subsequent procedures under sterile conditions in a Class II laminar flow biosafety cabinet.
5. Place a 0.45 μm cell strainer on a 50 mL conical tube.
6. Remove all muscle and connective tissue from femurs and tibiae using forceps and a scalpel (*see* **Note 8**).
7. Dip cleaned bones into 70 % ethanol for 10 s to sterilize, then place them in PBS to prevent drying prior to flushing (*see* **Note 9**).
8. Cut both ends of each bone using a scalpel and flush bone marrow with prewarmed (37 °C) cDMEM using a sterile 25-gauge syringe (*see* **Note 10**).
9. Pass the bone marrow suspension through the 0.45 μm cell strainer to remove debris.
10. Centrifuge the suspension at 400 g for 5 min
11. Discard the supernatant (*see* **Note 11**).
12. Resuspend the cell pellet in 3 mL RBC lysis buffer.
13. Incubate at room temperature for exactly 2 min
14. Neutralize RBC lysis by adding 7 mL cDMEM.
15. Centrifuge at 400 g for 5 min
16. Discard the supernatant.
17. Seed bone marrow cells into a 15 cm^2 dish containing 20 mL cDMEM supplemented with 20% (v/v) L929-CM (*see* **Notes 4 and 5**).
18. Maintain cultures for 7 days at 37 °C, 5 % CO_2.
19. On day 4, add an additional 20 mL cDMEM supplemented with 20% (v/v) L929-CM.

3.1.1 Plating of bone marrow-derived macrophages

1. On day 7, remove medium and wash BMDMs with 5 mL sterile Dulbecco's phosphate-buffered saline (DPBS).
2. Detach BMDMs by gently scraping with a sterile cell scraper in 5 mL DPBS.
3. Transfer the suspension to a 50 mL conical tube.
4. Rinse the dish with an additional 5 mL DPBS to collect residual cells and combine with the suspension.

5. Centrifuge at 400 g for 5 min
6. Discard the supernatant.
7. Resuspend the pellet in 5 mL prewarmed cDMEM supplemented with 20 % (v/v) L929-CM (see **Notes 4 and 5**).
8. Prepare a cell count: mix 10 μL of cell suspension with 80 μL DMEM and 10 μL trypan blue (*see* **Note 12**).
9. Count viable cells using a haemocytometer and adjust to 1×10^6 cells·mL^{-1} with cDMEM supplemented with 20% (v/v) L929-CM.
10. Plate 1 mL of cell suspension per well in a 12-well plate.
11. Incubate for 24 h at 37 °C, 5 % CO_2 prior to experimentation.

3.2 Generation of PMA-differentiated THP-1 macrophages

A timeline summary of the PMA differentiation protocol is presented in Fig. 1.

1. Culture THP-1 cells (maintain at 1-9 × 10^5 cells·mL^{-1}) in high-glucose complete RPMI 1640 medium (cRPMI) in T75 flasks.
2. Collect cells and medium into a 50 mL conical tube.
3. Centrifuge at 400 g for 5 min
4. Discard the supernatant.
5. Resuspend the pellet in 5 mL prewarmed cRPMI (*see* **Note 10**)
6. Prepare a cell count: mix 10 μL of suspension with 80 μL RPMI and 10 μL trypan blue (*see* **Note 12**).
7. Count viable cells with a haemocytometer and adjust to 1×10^6 cells·mL^{-1} using cRPMI.
8. Add PMA at 10 ng·mL^{-1} (1.6 mmol·L^{-1} final concentration) to the suspension (*see* **Note 13**).
9. Plate 2 mL of suspension per well in a 6-well plate (*see* **Note 14**).
10. Incubate for 48 h at 37 °C, 5 % CO_2 to induce differentiation.
11. Remove PMA-containing medium.
12. Gently wash away non-adherent cells with DPBS (*see* **Note 15**).
13. Replace with fresh cRPMI.
14. Rest cells for 5 h (optimum rest time is 5 h, but it can be up to 8 h) at 37 °C, 5 % CO_2 prior to experimentation.

3.3 Tolerance stimulation protocols (murine & human)

3.3.1 BMDM

1. Remove cDMEM supplemented with 20 % (v/v) L929-CM and wash cells.

2. For the unstimulated control (Unstim) and the non-tolerised control (NT), replace with fresh prewarmed cDMEM.
3. For tolerised wells (T), add cDMEM containing 100 ng·mL^{-1} LPS.
4. Incubate for 24 h at 37 °C, 5 % CO_2.
5. Wash cells with DPBS and replace medium with fresh cDMEM. Rest for 1 h.
6. After 1 h, stimulate NT and T wells with cDMEM containing 100 ng·mL^{-1} LPS for 3 h.
7. In the final 45 min, add ATP to a final concentration of 5 mmol·L^{-1} directly to each well.
8. Collect supernatants for later analysis.

3.3.2 PMA-differentiated macrophages

1. Following the generation of PMA-differentiated macrophages, and after the 5 h rest period has elapsed, replace the medium with cRPMI containing 10 ng·mL^{-1} LPS in the tolerized well (*see* **Note 16**).
2. Incubate for 24 h at 37 °C, 5 % CO_2.
3. Wash all wells (Unstim, NT, and T) with DPBS.
4. Replace medium with fresh prewarmed cRPMI and rest for 1 h.
5. After 1 h, stimulate NT and T wells with cRPMI containing 100 ng·mL^{-1} LPS for 3 h.
6. In the final 45 min, add ATP to a final concentration of 5 mmol·L^{-1}.
7. Collect supernatants for later analysis.

3.4 ELISA assay

A detailed ELISA protocol outlining the background and rationale of this assay is available in Methods in Molecular Biology Vol. 2283 (Nulty et al., 2021). A concise, practice-oriented version is provided below.

3.4.1 Assay procedure

1. Dilute the capture antibody in 1 × coating buffer.
2. Add 50 μL of capture antibody solution to each well of a 96-well high-binding plate.
3. Seal and incubate overnight (16–18 h) at 2–8 °C.
4. Wash the plate 4 times with 300 μL wash buffer per well. Blot residual liquid by tapping the plate upside down on absorbent paper. Perform all subsequent washes similarly.
5. Add 100 μL of 1 × assay diluent to each well to block non-specific binding.
6. Seal and incubate for 1 h at room temperature (RT) on a plate shaker.

7. Wash the plate 4 times with wash buffer.
8. Add 50 μL of standards or samples (diluted as necessary in 1 × assay diluent) to the appropriate wells (*see* **Note 17**).
9. Seal and incubate for 2 h at RT with shaking.
10. Wash the plate 4 times with wash buffer.
11. Add 50 μL of diluted detection antibody to each well.
12. Seal and incubate for 1 h at RT with shaking.
13. Wash the plate 4 times with wash buffer.
14. Add 50 μL of diluted avidin-HRP solution to each well.
15. Seal and incubate for 30 min at RT with shaking.
16. Wash the plate 5 times with wash buffer; during the final wash, soak wells for 30–60 s to minimize background.
17. Add 50 μL of freshly mixed TMB substrate (A + B, 1:1) to each well.
18. Incubate for 20 min in the dark. Do not seal the plate.
19. Stop the reaction by adding 50 μL of stop solution to each well.
20. Read absorbance at 450 nm and 570 nm within 15 min

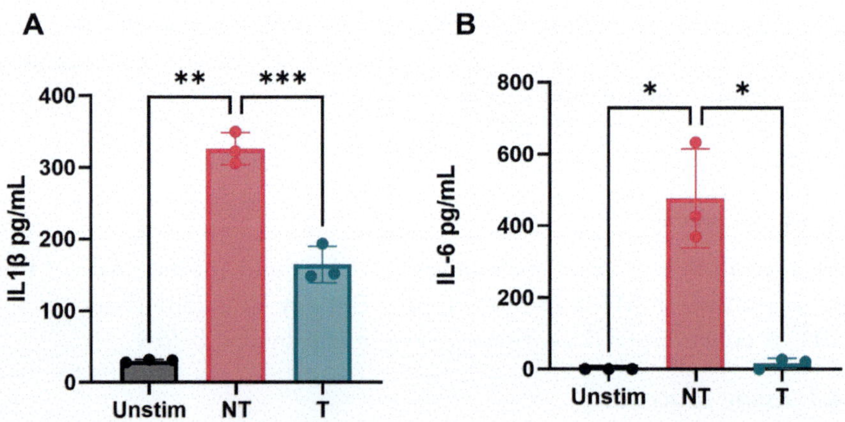

Fig. 2 Induction of tolerance phenotype in human PMA differentiated macrophages. THP-1 cells were seeded at 1×10^6 cells/mL in 6-well plates and differentiated into macrophages with PMA (1.6 mM) for 48 h, followed by a 5 h rest period. PMA-differentiated macrophages were subsequently stimulated with LPS (10 ng/mL) or culture media for 24 h to induce the tolerised (T) or non-tolerised (NT) phenotype, respectively. Following cell washing and a 1 h rest period, cells were re-stimulated with LPS (10 ng/mL, 3 h 45 min) and ATP (5 mM, 45 min) to activate the canonical inflammasome, as shown in Fig. 1C. Supernatants were collected and analysed by ELISA for (A) IL-1β and (B) IL–6 concentration. Values represent the mean ± SD of n = 3. Statistical analysis was performed using one-way ANOVA (Tukey post-hoc test), *P < 0.05, **P < 0.01, ***P < 0.001.

3.4.2 Data analysis

1. Export 450 nm and 570 nm readings into Excel.
2. Subtract 570 nm values from 450 nm values (*see* **Note 18**).
3. Calculate the mean of duplicate values.
4. Correct for blank absorbance by subtracting the blank from all standards and samples.
5. In Prism, open a new XY graph.
6. Enter standard concentrations (X) and corresponding standard absorbances (Y).
7. Paste blank-corrected mean sample absorbances beneath the standards and use the interpolation function to calculate cytokine concentrations (*see* **Note 19**).
8. Create column or grouped graphs as appropriate (use grouped graphs for genotype comparisons).
9. Repeat the experiment (n = 3) for statistical validity; independent samples can be run on the same ELISA plate.
10. For column analyses, perform a one-way ANOVA with Tukey correction; for grouped analyses, perform a two-way ANOVA with Sidak correction.

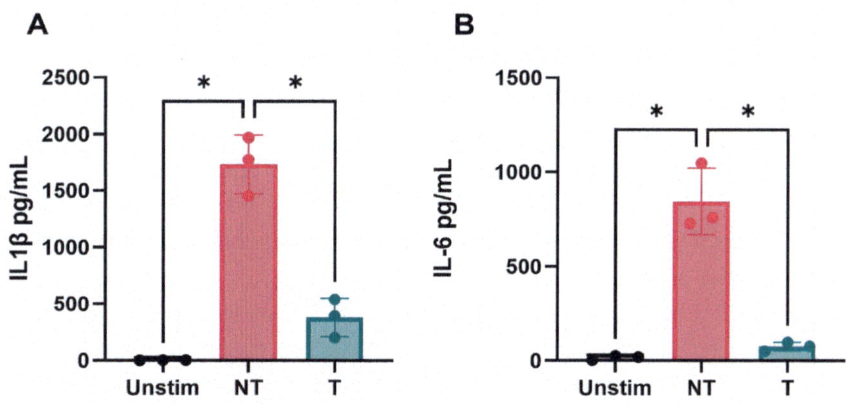

Fig. 3 Induction of tolerance phenotype in murine bone marrow derived macrophage cells (BMDM). Murine BMDM were plated 24 h prior to treatment in a 12 well plate, at 1×10^6 cells/mL. BMDM were stimulated with 100 ng/mL LPS for 24 h to induce tolerised phenotype (T) or left untreated for non-tolerised phenotype (NT). Following cell washing and a 1 h rest period, cells were stimulated with 100 ng/mL LPS for 3 h and 45 mins ATP (5 mM) to activate the canonical inflammasome. Supernatants were collected and analysed by ELISA, with IL-1β (A) concentration used as a measure of NLRP3 inflammasome activation/tolerance and IL-6 (B) used as a measurement of NFκB activation/tolerance. Values represent the mean ± SD of n = 3. Statistical analysis was performed using one-way ANOVA (Tukey post-hoc test), $^*P < 0.05$.

3.4.3 Interpreting results

The protocol, summarised in Fig. 1, reliably establishes endotoxin tolerance models in macrophages derived from wild-type THP-1 (Fig. 2) and BMDM cells (Fig. 3). Upon stimulation with LPS and ATP, to activate the canonical inflammasome, non-tolerised macrophages produce significantly higher concentrations of pro-inflammatory cytokines IL-1β (Figs. 2A and 3A), and IL-6 (Figs. 2B and 3B), when compared to tolerised and unstimulated macrophages ($p < 0.05$), in both human and murine systems. This provides a robust framework for assessing whether tolerance is statistically altered in murine and human macrophages. These protocols can be used and adapted to interrogate signalling mediators that regulate LPS tolerance, by examining the effects of genetic manipulation, or the use of inhibitors or antagonists on the tolerance phenotype.

4. Notes

1. Alternative suppliers and products may be used.
2. BMDM cells were isolated from femurs and tibiae of adult C57BL/6 J mice (> 8 weeks). Femurs and tibiae can be extracted from mice bred in-house within an institute's animal facility, or legs can be obtained from commercial suppliers e.g. /http://www.jax.org.
3. As this protocol uses endotoxin to stimulate cells, it is essential that each FBS batch/lot from suppliers are tested for endotoxin levels. Only FBS with endotoxin concentrations < 0.2 $EU{\cdot}mL^{-1}$ should be used. An example of a supplier which routinely test each FBS lot for levels of endotoxin is Merck.
4. The L929 cell line is an immortalized mouse fibroblast line derived from subcutaneous connective, areolar, and adipose tissue of a 100-day-old male C3H/An mouse. L929 cells constitutively express M-CSF and other stimulatory factors. As such, supplementation with 20 % L929 conditioned medium (CM) is commonly used as a source of M-CSF when differentiating BMDM. Recombinant M-CSF (20 ng/mL) can also be used to generate macrophages from BMDM (de Brito Monteiro et al., 2020).
5. L929 CM is produced by culturing L929 cells in complete DMEM supplemented with 10 % FBS and 1 % PenStrep, at a density of 20×10^6 cells in a T175 cm^2 flask with 40 mL (5×10^5 $cells{\cdot}mL^{-1}$)

complete DMEM for 10 days. After this time, supernatant media from the L929 cell culture is filtered through 0.22 μm sterile syringe filter and aliquoted before storing at −20 °C

6. Animal husbandry ethics legislation varies by region. It is important to be familiar with and follow the local ethical practices on animal euthanasia and husbandry.
7. The simplest way to excise the leg bone is to cut the tissue surrounding the hip joint (taking care not to damage the bone), then gently dislocate the femur by pulling on the leg and cutting the remaining tissue attaching the leg to the body.
8. An efficient way to clean bones is to combine stripping tissue away with forceps and cutting residual tissue with a scalpel.
9. A 24-well plate can be used to hold 70 % ethanol and PBS for dipping and keeping bones moist while working.
10. When working with living cells, media must be pre-warmed to 37 °C in a water bath before being added to cells.
11. All cell culture reagents and media that have been in contact with biological material must be placed in Virkon for sterilisation before disposal into biological waste.
12. Trypan blue is an azo dye used to assess cell viability. It selectively stains dead cells blue, as it cannot cross the intact membranes of viable cells.
13. High concentrations of PMA (≥50 ng·mL^{-1}) can partially activate THP-1 macrophages, which may alter their response to subsequent LPS stimulation. This lower concentration has been optimised in our laboratory.
14. PMA-differentiated macrophages are smaller than murine BMDMs and therefore yield lower protein concentrations in lysates. If techniques requiring high protein concentration, such as western blotting, are required for analysis, it is recommended to scale up to a 6-well plate (as described in this protocol).
15. Non-adherent cells have not differentiated properly into macrophages and removed during wash steps.
16. THP-1 cells can spontaneously activate their inflammasome in response to high concentrations of LPS (Park et al., 2007). The 10 ng/mL LPS concentration used to induce tolerance in this protocol does not induce spontaneous inflammasome activation in THP-1 macrophages.

17. The need to dilute samples is dependent on the concentration of the analyte in the sample and the assay kit sensitivity. Dilution optimisations may be necessary.
18. Subtracting absorbance at 570 nm from 450 nm helps improve specificity by removing background interference caused by plate artifacts, bubbles, or non-specific absorbance.
19. Choose the most appropriate interpolation method depending on whether your standard curve fits a linear or sigmoidal model. The R^2 value indicates the goodness of fit, with values closer to 1 representing a better fit. This can help determine the best regression for your data.

5. Concluding remarks

As detailed in the introductory section, endotoxin tolerance in macrophages is essential for the maintenance of cell survival and function, especially in environments with repeated/prolonged LPS exposure, such as mucosal tissues (Roszak et al., 2025). However, when LPS tolerance becomes dysregulated or prolonged, it may contribute to adverse outcomes, including immune suppression and increased susceptibility to secondary infections, as observed in sepsis (Venet et al., 2021). Conversely, in autoinflammatory conditions, the inability of macrophages to establish tolerance to environmental LPS exposure can be equally detrimental, leading to excessive inflammation and cell death (Larsson et al., 2022). Therefore, a detailed understanding of tolerance mechanisms in macrophages is necessary to improve patient outcomes.

Murine models and in vitro experiments using mouse-derived cells are commonly employed to study LPS tolerance, as genetically modified animals allow interrogation of the genes and proteins involved in this process. Using this approach can be helpful in improving mechanistic knowledge as well identifying new genes involved in the process. The murine tolerance stimulation protocol presented here was developed by the Artyomov group (Bambouskova et al., 2021), and similar approaches have been employed by various other groups (Boonmee et al., 2024; Gillen et al., 2021; Naler et al., 2022). While mouse models are invaluable for early mechanistic studies, human systems are also crucial, particularly for genes that are differentially expressed between species (Silswal et al., 2017; Wang et al., 2022). For example, the LPS-responsive gene encoding for inducible nitric oxide synthase (*NOS2*) is expressed at different levels in human and murine macrophages, highlighting the importance of using both models when investigating tolerance mechanisms (Gilbertson et al., 2022; Schroder et al., 2012).

Primary human macrophages, monocyte-derived macrophages (MDMs), derived from healthy patients and those with inflammation-associated disease can also serve as a useful human model for tolerance. However, the genetic modification of primary cells is technically challenging and limits their utility. In contrast, THP-1 cells are an immortalised human monocytic cell line that can be genetically manipulated. Although PMA-differentiated THP-1 macrophages are often hyperresponsive to inflammatory stimuli such as LPS, restricting their broader use, this protocol is an optimised tolerance protocol in PMA-differentiated THP-1 cells that accounts for both PMA-induced activation and LPS hyperresponsiveness (Maeß et al., 2014; Park et al., 2007). Consequently, this model provides a viable model, which can be readily genetically manipulated, to study endotoxin tolerance.

In this study, endotoxin tolerance was assessed using ELISA-based quantification of cytokine secretion, a widely adopted functional readout in both murine and human macrophage models. ELISA offers high sensitivity and specificity for individual cytokines, relatively low cost per sample, and straightforward implementation, making it well suited for scalable studies. Importantly, this approach directly captures one of the defining functional features of endotoxin tolerance, attenuated cytokine production upon secondary LPS challenge, which is equally applicable to murine- and human-derived macrophages.

Alternative methods used to assess tolerance include transcriptomic approaches, epigenetic profiling (Ghiboub et al., 2020; Naler et al., 2022; Sedivy-Haley et al., 2022), flow cytometry–based intracellular cytokine staining (Davtyan et al., 2006), and multiplex cytokine assays (Koch et al., 2017). Transcriptomic and epigenetic techniques provide high sensitivity, enabling identification of tolerance-associated gene networks and chromatin remodelling events in both species. However, these methods are comparatively costly and require specialized expertise in handling big datasets (Isaic et al., 2025; Van den Berge et al., 2022). Flow cytometry allows single-cell resolution and phenotypic stratification, but has increased technical complexity when compared to ELISA, and can be limited by fluorophore compatibility (Wilson et al., 2018). Multiplex cytokine assays have increased throughput relative to ELISA, which can be extremely useful when sample volume is limited. However, use of multiplex immunoassays can often be limited by their relative high cost, in addition to issues relating to immunoassay quality control and standardization, due to the complexity of testing multiple analytes simultaneously (Tighe et al., 2015).

Overall, ELISA-based cytokine evaluation represents a robust, cost-effective, and scalable approach for functional assessment of endotoxin tolerance in both murine and human macrophage models. While this approach lacks the global mechanistic screening ability of transcriptomic or epigenomic analyses, the single cell resolution of flow cytometry and the increased analyte coverage of multiplex assays, its ease of use and reproducibility make it particularly suitable for comparative studies and initial screening efforts. Thus, this method provides a practical and informative framework for studying endotoxin tolerance and for identifying conserved and divergent regulatory mechanisms between murine and human macrophages.

Acknowledgements

This work is supported by an Enterprise Ireland Commercialisation Fund (CF-2020–1571-A), co-funded by the European Regional Development Fund under Ireland's European Structural and Investment Funds Programme. Mathew Choaji is supported by the Project 101119873 NO-CANCER-NET, Horizon Europe MSCA DN (under grant agreement HORIZON-MSCA-2022-DN-01 project No 101119873).

References

Agnew, A., Nulty, C., & Creagh, E. M. (2021). Regulation, activation and function of caspase-11 during health and disease. *International Journal of Molecular Sciences, 22*.

Bambouskova, M., Potuckova, L., Paulenda, T., Kerndl, M., Mogilenko, D. A., Lizotte, K., ... Artyomov, M. N. (2021). Itaconate confers tolerance to late NLRP3 inflammasome activation. *Cell Reports, 34*, 108756.

Boonmee, A., Benjaskulluecha, S., Kueanjinda, P., Wongprom, B., Pattarakankul, T., Sri-Ngern-Ngam, K., ... Palaga, T. (2024). A polycomb group protein EED epigenetically regulates responses in lipopolysaccharide tolerized macrophages. *Epigenetics Chromatin, 17*, 36.

Chen, S., Saeed, A., Liu, Q., Jiang, Q., Xu, H., Xiao, G. G., ... Duo, Y. (2023). Macrophages in immunoregulation and therapeutics. *Signal Transduction and Targeted Therapy, 8*, 207.

Davtyan, T. K., Hakopyan, G. S., Avetisyan, S. A., & Mkrtchyan, N. R. (2006). Impaired endotoxin tolerance induction in patients with familial Mediterranean fever. *Pathobiology: Journal of Immunopathology, Molecular and Cellular Biology, 73*, 26–39.

de Brito Monteiro, L., Davanzo, G. G., de Aguiar, C. F., Corrêa da Silva, F., Andrade, J. R., Campos Codo, A., ... Moraes-Vieira, P. M. (2020). M-CSF- and L929-derived macrophages present distinct metabolic profiles with similar inflammatory outcomes. *Immunobiology, 225*, 151935.

Ghiboub, M., Zhao, J., Li Yim, A. Y. F., Schilderink, R., Verseijden, C., van Hamersveld, P. H. P., ... de Jonge, W. J. (2020). HDAC3 mediates the inflammatory response and LPS tolerance in human monocytes and macrophages. *Frontiers in Immunology, 11*, 550769.

Gilbertson, S. E., Walter, H. C., Gardner, K., Wren, S. N., Vahedi, G., & Weinmann, A. S. (2022). Topologically associating domains are disrupted by evolutionary genome rearrangements forming species-specific enhancer connections in mice and humans. *Cell Reports, 39*, 110769.

Gillen, J., Ondee, T., Gurusamy, D., Issara-Amphorn, J., Manes, N. P., Yoon, S. H., ... Nita-Lazar, A. (2021). LPS tolerance inhibits cellular respiration and induces global changes in the macrophage secretome. *Biomolecules, 11*.

Hoshino, K., Takeuchi, O., Kawai, T., Sanjo, H., Ogawa, T., Takeda, Y., ... Akira, S. (1999). Cutting edge: Toll-like receptor 4 (TLR4)-deficient mice are hyporesponsive to lipopolysaccharide: evidence for TLR4 as the Lps gene product. *Journal of Immunology, 162*, 3749–3752.

Isaic, A., Motofelea, N., Hoinoiu, T., Motofelea, A. C., Leancu, I. C., Stan, E., ... Crintea, A. (2025). Next-generation sequencing: A review of its transformative impact on cancer diagnosis, treatment, and resistance management. *Diagnostics (Basel), 15*.

Kawai, T., Ikegawa, M., Ori, D., & Akira, S. (2024). Decoding Toll-like receptors: Recent insights and perspectives in innate immunity. *Immunity, 57*, 649–673.

Koch, R. M., Kox, M., Thijs, E. J. M., Rahamat-Langendoen, J. C., van de Veerdonk, F. L., Gerretsen, J., ... Pickkers, P. (2017). Development of endotoxin tolerance does not influence the response to a challenge with the mucosal live-attenuated influenza vaccine in humans in vivo. *Frontiers in Immunology, 8*, 1600.

Larsson, L., Garaicoa-Pazmino, C., Asa'ad, F., & Castilho, R. M. (2022). Understanding the role of endotoxin tolerance in chronic inflammatory conditions and periodontal disease. *Journal of Clinical Periodontology, 49*, 270–279.

Li, H., Breedijk, A., Dietrich, N., Nitschke, K., Jarczyk, J., Nuhn, P., ... Hauske, S. (2023). Lipopolysaccharide tolerance in human primary monocytes and polarized macrophages. *International Journal of Molecular Sciences, 24*.

Luo, R., Yao, Y., Chen, Z., & Sun, X. (2025). An examination of the LPS-TLR4 immune response through the analysis of molecular structures and protein-protein interactions. *Cell Communication and Signaling: CCS, 23*, 142.

Maeß, M. B., Wittig, B., Cignarella, A., & Lorkowski, S. (2014). Reduced PMA enhances the responsiveness of transfected THP-1 macrophages to polarizing stimuli. *Journal of Immunological Methods, 402*, 76–81.

Mages, J., Dietrich, H., & Lang, R. (2007). A genome-wide analysis of LPS tolerance in macrophages. *Immunobiology, 212*, 723–737.

Naler, L. B., Hsieh, Y. P., Geng, S., Zhou, Z., Li, L., & Lu, C. (2022). Epigenomic and transcriptomic analyses reveal differences between low-grade inflammation and severe exhaustion in LPS-challenged murine monocytes. *Communications Biology, 5*, 102.

Nulty, C., Barber, G., Flis, E., & Creagh, E. M. (2021). Evaluating host responses to helicobacter pylori using ELISA and western blot. *Methods in Molecular Biology, 2283*, 175–190.

Park, E. K., Jung, H. S., Yang, H. I., Yoo, M. C., Kim, C., & Kim, K. S. (2007). Optimized THP-1 differentiation is required for the detection of responses to weak stimuli. *Inflammation Research, 56*, 45–50.

Rhee, S. H., & Hwang, D. (2000). Murine TOLL-like receptor 4 confers lipopolysaccharide responsiveness as determined by activation of NF kappa B and expression of the inducible cyclooxygenase. *The Journal of Biological Chemistry, 275*, 34035–34040.

Roszak, K., Roy, K., Sobocińska, J., Spisz, P., Jędrzejewski, T., & Wrotek, S. (2025). Endotoxin's impact on organism: From immune activation to tolerance and beyond. *Journal of Clinical Medicine, 14*.

Schroder, K., Irvine, K. M., Taylor, M. S., Bokil, N. J., Le Cao, K. A., Masterman, K. A., ... Sweet, M. J. (2012). Conservation and divergence in Toll-like receptor 4-regulated gene expression in primary human versus mouse macrophages. *Proceedings of the National Academy of Sciences of the United States of America, 109*, E944–E953.

Sedivy-Haley, K., Blimkie, T., Falsafi, R., Lee, A. H., & Hancock, R. E. W. (2022). A transcriptomic analysis of the effects of macrophage polarization and endotoxin tolerance on the response to Salmonella. *PLoS One, 17*, e0276010.

Seeley, J. J., & Ghosh, S. (2017). Molecular mechanisms of innate memory and tolerance to LPS. *Journal of Leukocyte Biology, 101*, 107–119.

Silswal, N., Reis, J., Qureshi, A. A., Papasian, C., & Qureshi, N. (2017). Of mice and men: Proteasome's role in LPS-induced inflammation and tolerance. *Shock (Augusta, Ga.), 47*, 445–454.

Smith, A. P., & Creagh, E. M. (2022). Caspase-4 and -5 biology in the pathogenesis of inflammatory bowel disease. *Frontiers in Pharmacology, 13*, 919567.

Tighe, P. J., Ryder, R. R., Todd, I., & Fairclough, L. C. (2015). ELISA in the multiplex era: Potentials and pitfalls. *Proteomics. Clinical Applications, 9*, 406–422.

Van den Berge, K., Chou, H. J., Roux de Bézieux, H., Street, K., Risso, D., Ngai, J., & Dudoit, S. (2022). Normalization benchmark of ATAC-seq datasets shows the importance of accounting for GC-content effects. *Cell Reports Methods, 2*, 100321.

Venet, F., Demaret, J., Gossez, M., & Monneret, G. (2021). Myeloid cells in sepsis-acquired immunodeficiency. *Annals of the New York Academy of Sciences, 1499*, 3–17.

Wang, N., Lu, Y., Zheng, J., & Liu, X. (2022). Of mice and men: Laboratory murine models for recapitulating the immunosuppression of human sepsis. *Frontiers in Immunology, 13*, 956448.

Widdrington, J. D., Gomez-Duran, A., Pyle, A., Ruchaud-Sparagano, M. H., Scott, J., Baudouin, S. V., ... Simpson, A. J. (2018). Exposure of monocytic cells to lipopolysaccharide induces coordinated endotoxin tolerance, mitochondrial biogenesis, mitophagy, and antioxidant defenses. *Frontiers in Immunology, 9*, 2217.

Wilson, A. C., Moutsatsos, I. K., Yu, G., Pineda, J. J., Feng, Y., & Auld, D. S. (2018). A scalable pipeline for high-throughput flow cytometry. *SLAS Discovery, 23*, 708–718.

CHAPTER ELEVEN

Assessment of microglial responsiveness after irradiation in mouse organotypic hippocampus or patient-derived tissue cultures

Bahar Djouiai[a,f,1], Felicitas Rapp[b,f,1], Franziska Köhler[a,b], Frank Gaunitz[c], Steffen Rosahl[d], Rüdiger Gerlach[d], Faramarz Dehghani[e], Ingo Bechmann[a], Katharina Heß[g], Karsten Winter[a], and Sonja Kallendrusch[a,g,*]

[a]Institute of Anatomy, University of Leipzig, Leipzig, Germany
[b]Department of Hematology and Oncology, Medical Faculty Mannheim, Heidelberg University, Mannheim, Germany
[c]Department of Neurosurgery, University Hospital Leipzig, Leipzig, Germany
[d]Department of Neurosurgery, HELIOS Hospital Erfurt, Erfurt, Germany
[e]Institute of Anatomy, University of Halle, Halle, Germany
[f]GSI Helmholtzzentrum für Schwerionenforschung, Biophysics Department, Darmstadt, Germany
[g]Institute of Clinical Research and Systems Medicine, HMU Health and Medical University Potsdam, Potsdam, Germany
*Corresponding author. e-mail address: sonja.kallendrusch@hmu-potsdam.de

Contents

[1] Contributed equally.

Methods in Cell Biology, Volume 210
ISSN 0091-679X, https://doi.org/10.1016/bs.mcb.2026.07.001

Abstract

Understanding radiation-induced tissue modifications and cell death within the complex cellular architecture of the brain requires experimental systems that preserve tissue integrity and intercellular crosstalk. At the same time, microglial cells as primary regulators of primary and secondary inflammation in the brain are difficult to study in single cell cultures and it is crucial to investigate their behaviour in their complex tissue microenvironment. Rodent organotypic hippocampal slice cultures (OHSCs) and human patient-derived brain tumor cultures (PDTCs) offer physiologically relevant platforms for investigating the differential effects of X-ray and heavy-ion irradiation on glial as well as immune cell populations. Both models retain native cytoarchitecture, extracellular matrix composition, and functional cell–cell interactions, making them uniquely suited to study the spatial and temporal dynamics of inflammation or cell death in a multicellular context. Here, we present a standardized methodology for preparing, irradiating, and assessing PDTCs, employing multimodal readouts as cytokine measurement of the supernatant, cell death, microglia morphology and function. Novel molecular analyses are possible upon certain considerations and may alter our understanding of species differences and help to distinguish cell death pathways to attribute damage to specific cell types. Standardization of these protocols across model systems is essential for generating reproducible, comparative data while preserving the inherent complexity that underlies intercellular communication and coordinated responses to injury. This approach enables translationally relevant insight into the mechanisms of radiation-induced brain injury to develop strategies for neuroprotection and treatment in clinical settings.

1. Introduction

The brain has long been considered an immunologically privileged organ, reflecting both its restricted exchange with the periphery via the blood–brain barrier (BBB) and the persistent nature of neurons, which favor repair over replacement or cell death (Carson et al., 2006; Daneman & Prat, 2015). Yet, following injury, neurons and glial cells release cytokines to initiate repair responses. These signals, together with astrocyte and microglial activity and, in the case of a compromised BBB, the recruitment of peripheral immune cells, orchestrate the regulation of damage-associated molecular patterns (DAMPs) and shape the tissue response (Gadani et al., 2015). Attempts to broadly suppress neuroinflammation have largely failed (Bourne et al., 2025). Systemic immunosuppression can be counterproductive, as it inhibits microglial activation, a process essential for the resolution of inflammation (Schwartz & Baruch, 2014). These failures highlight that CNS repair is dynamic and temporally regulated, requiring interventions that are both context- and time-specific to avoid secondary damage and maladaptive long-term repair (Biswas, 2023; Kallendrusch et al., 2012, 2013; Preissler et al., 2015; Russo & McGavern, 2016).

Brain injury typically provokes a multiphase response. Early glial surveillance shifts to phagocytosis, which may then progress into pathological activation, senescence, or dysfunctional cell death, thereby sustaining neuroinflammation and tissue degeneration. This cascade is largely mediated by microglia, the resident macrophages of the brain (Prinz et al., 2021; Wang et al., 2024). Through their pattern-recognition receptors (PRRs), microglia sense host-derived molecules, triggering signalling cascades and transcriptional programs that culminate in inflammatory mediator production (Murray et al., 2014). Microglia adopt diverse phenotypes: pro-inflammatory states associated with neurodegeneration, and alternative states with anti-inflammatory or neuroprotective properties, further distinguished by their secretory profiles (Voshart et al., 2024; Wang et al., 2024).

Studying microglia in vitro presents challenges. These cells rapidly adapt to changes in their microenvironment and are highly sensitive to dissociation and sorting, which can induce artificial activation signatures (e.g., immediate-early gene expression, stress responses, or immune markers) (Warden et al., 2023). This underscores the importance of stringent protocols and advanced experimental systems for accurately modelling human microglial biology (Mulenge et al., 2024; Schafer et al., 2023). Tissue cultures derived from murine and human sources have proven valuable for investigating CNS pathologies, although translating findings from animal models remains limited by pronounced species differences (Kallendrusch et al., 2012; Merz et al., 2010; Patties et al., 2019; Voshart et al., 2024; Warden et al., 2023).

Radiation, either from treatment scenarios or other sources like space flights, has emerged as a double-edged factor in neurobiology. Low-dose ionizing radiation has been proposed as a treatment for neuronal injury and stroke, aiming to attenuate secondary damage mediated by microglia and macrophages (Au et al., 2024). Conversely, radiation exposure, including X-rays, high-energy protons or cosmic radiation during space missions, poses risks that demand deeper investigation (Fogtman et al., 2023; Merz et al., 2013; Salem et al., 2024; Tinganelli et al., 2021). Regardless of type, radiation induces DNA damage, mitochondrial ROS production, and DAMP release. These events rapidly shift microglia from surveillant to amoeboid phenotypes, increasing phagocytic activity within minutes to hours (Fig. 1). Activated microglia secrete IL-6, TNF, and

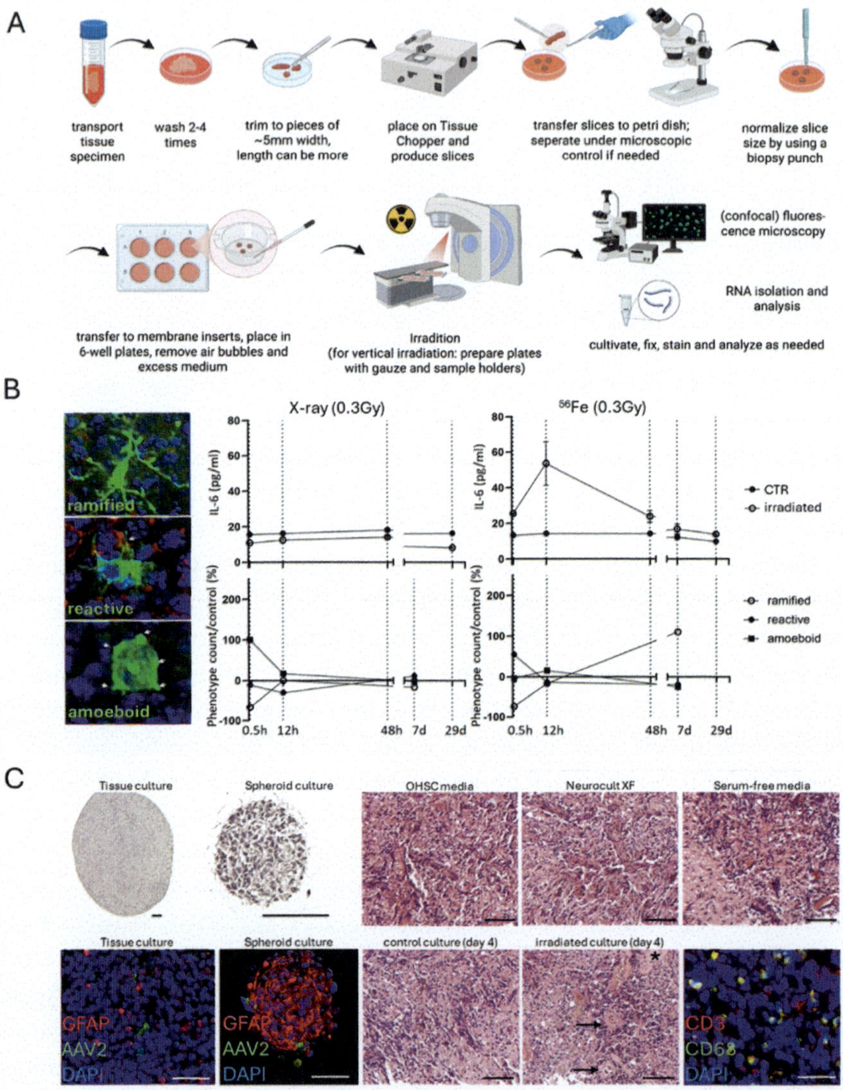

Fig. 1 Radiation effects in tissue cultures. (A) *Schematic presentation of the procedure for human tissue culture.* After tissue processing by tissue chopper or a vibratome, tissue is cultured at an air-liquid interface necessary to guarantee oxygen as well as nutritional supply. Irradiation with photons or (heavy) ions and subsequent analysis can be performed as needed. Created by Biorender. (B) *Microglial characterisation in OHSC and IL-6 measurement in the supernatant after low-dose irradiation with X-ray and ^{56}Fe heavy ions.* Representative immunofluorescence images of three microglial phenotypes are shown (left panel, IB4 positive microglia in green, nuclei blue, white arrows point to retracted (amoeboid) or debris engulfing (reactive) branches). The effect of low-dose-x-ray radiation and ^{56}Fe heavy ion radiation over a time period

IL-1β, undergo morphological reorganization, and may develop heightened sensitivity to subsequent insults, thereby potentially prolonging inflammation through secondary phases of activation (Voshart et al., 2024). Persistent activation of the STING pathway and sustained inflammatory signalling have been linked to cognitive decline and neurodegeneration, particularly in patients undergoing brain tumor therapy (Constanzo et al., 2021; Gulen et al., 2023). However, therapeutic strategies and risk assessments remain insufficiently defined in these contexts, underscoring the need for robust experimental models (Au et al., 2024; Fogtman et al., 2023; Lawrie et al., 2019; Merz et al., 2013; Salem et al., 2024; Tinganelli et al., 2021; Toprani et al., 2024; Voshart et al., 2024). Still, species differences further complicate translation.

To overcome these challenges, we describe a system for patient-derived tissue cultures of brain tumors obtained during surgical resection and how to irradiate this system model. This platform further enables in vivo imaging, spatial transcriptomics, morphological characterization, and functional assays. By maintaining the native cellular and immunological context, such cultures represent a powerful model to address species-specific limitations and have already proven valuable for investigating Tumor Treating Fields as well as drug and radiation susceptibility (Leyh et al., 2021; Merz et al., 2010, 2013; Nickl et al., 2022). It is also possible to use murine organotypic hippocampal slice cultures (OHSC)

---

from 0.5 h (h) to 29 days (d) is shown (right panel). IL-6 concentration is depicted in the upper graphs (left: Xray, right: Iron (^{56}Fe) ions; n = 3). Microglial phenotype alterations are shown by the mean change, relative to their respective OHSC controls (lower graphs left: Xray, right: Iron (^{56}Fe) ions; shown is the mean of n = 3). (C) *Patient-derived glioblastoma tissue samples or biopsies are used for tissue culture as well as spheroid culture.* The difference in size of patient-derived tissue cultures (left, Bar = 100 μm) to patient-derived spheroids (right, Bar = 100 μm) created of the same patient-tissue is given. In the lower panel adeno-associated viral vector 2 (AAV2) transduction was performed in both model systems to test tissue penetration (PDTC (left) and spheroid (right) were paraffin embedded prior staining, Bar = 50 μm). Different culture media were tested to determine stability of the tissue composition. A good morphological maintenance was observed with media adapted from rodent organotypic hippocampal slice cultures as well as in the commercially available serum-free Neurocult XF proliferation medium, used also for spheroid cultivation. Other media tested, without serum (B27 or horse sera) addition does not maintain the native structure of the tissue. The non-irradiated control condition is shown, while the respective PDTC was irradiated (4 Gy) and demonstrates necrosis (asterix) as well as sclerotic alterations of blood vessels (arrows). Maintenance of CD3 positive T-cells is shown in fluorescence after four days of cultivation.

to study basic reactions of brain tissue to irradiation or treatments. Generation of OHSC is published and widely used, please refer to validated protocols (Stoppini et al., 1991).

2. Materials

2.1 Ethics

This research was conducted in accordance with the ethical principles of the Declaration of Helsinki (World Medical Association World Medical Association, 2024). Handling of human tissue requires approval from the local ethics committee. Only tissue that is not needed for routine diagnostic purposes is used in this protocol. Prior to each experiment, written informed consent must be obtained from the donors, and proper documentation must be ensured.

2.2 Disposables

General disposables.

1. Petri dishes 100 mm (#664160, Greiner)
2. Petri dishes 60 mm (#93060, TPP)
3. Pasteur pipette 150 mm (#7691060, Labsolute) (Note 1)
4. Biopsy punch 2–3 mm (#48301, B.Braun)
5. 6-well Plate (# 92006, TPP)
6. Millicell ® Inserts, 0,4 µm PFTE, 33 mm (#PICM0RG50, Merck Millipore) (Note2)
   - filter paper (#205005, Macherey Nagel Ø 55 mm)
   - razor blade
   - scalpel blade (B.Braun No. 11 and No. 15)
   - Histoacryl ® (tissue adhesive) (#1050052, B.Braun)
   - Gaze tissue Gazin® (#PZN2386908, Lohmann and Rauscher)

Instruments:

- Forceps (rounded, straight, curved)
- Smaller forceps (pointed, rounded, straight, curved)
- Scalpel for blades No11 and No15

Reagents: (Note 3).

- Minimal Essential Medium (#21090-022, with Earle's Salts, Gibco)
- Hank's Balanced Salt Solution (#24020, Gibco)
- Normal horse serum (#SH3007 4.03, GE Healthcare)

- GlutaMAX (#35050-061, Gibco)
- Neurocult (#05761, NC-XF human (Basal and Proliferation supplement), Stemcell Technologies)
- rh EGF (#78006, Stemcell Technologies)
- rh FGF (#78003, Stemcell Technologies)
- Heparin Solution (#07980, Stemcell Technologies)
- RNA-stabilization reagent (RNAlater) (#AM7020, Invitrogen)

Optional:
- Penicillin/Streptomycin (#15140-122, Gibco)

2.3 Equipment
- Tissue chopper (Cavey Laborator Engineering Co. Ltd., McIlwain Tissue Chopper, Model TC752) (Note 4)
- Stereomicroscope (EVIDENT Europe GmbH (former Olympus), Model SZX16)
- Biological safety cabinet (Fisher Scientific GmbH, Model Herasafe 2030i)
- Laminar flow bench (Nordic Labtech AB, Model nSAFE Horizontal)
- X-ray irradiator (e.g., XRAD 320, Precision X-Ray)
- Facility to deliver ion irradiation (e.g., GSI Helmholtz Centre, HIT Heidelberg)
- Dosimetry equipment (ionization chambers, film badges)
- Slidescanner Zeiss Axioscan 7 (Carl Zeiss Microscopy GmbH, Jena, Germany)

2.4 Software
- ZEN (Carl Zeiss Microscopy GmbH, Jena, Germany)
- ImageJ (http://imagej.hih.gov/ij)
- Qupath,
- Mathematica (Wolfram Research Inc., Champaign, IL, USA)
- Graphpad Prism

3. Methods

3.1 Media recipes
- Preparation media
  - o Minimal Essential Medium (#21090-022, with Earle's Salts, Gibco)
  - o Add 1–5 % antibiotics (Penicillin/ Streptomycin, optional).
  - o For transportation use 10 ml for 50 ml tubes, 2–3 ml for 15 ml tubes.

- Cultivation media
  - o MEM (Gibco Ref. 32360-026)
    +25 % HBSS $CaCl_2$ $MgCl_2$
    +5 % Normal Horse Serum
    +1 % GlutaMAX 200 mM
    +1 % of 45 % Glucose solution
  - o Neurocult-XF Basal Medium
    +10 %Neurocult-XF Proliferation Supplement.
    +0.2 %rh EGF.
    +0.1 %rh FGF.
    +0.1 %Heparin solution.
    (optional) Penicillin/Streptomycin + 1 % Pen/Strep.

3.2 Tissue acquisition

- Remove the tissue from the transportation tube filled with preparation media (15 ml tubes or 50 ml tubes). (Note 5)
- Wash 2–4 times in preparation media.
- Section the tissue, depending on size, to approx. 5 × 7 mm pieces. (Note 6)
- Transfer into clean petri dish (60−100 mm diameter) in preparation media.

3.3 Tissue processing

1. Cover the tissue chopper platform with 3−4 filter papers.
2. Mount the razor blade onto the tissue chopper. Adjust blade without tissue to the stage.
3. Adjust/check settings to 350 μm slicing steps and cutting speed (Note 7).
4. Wet the filter paper, but not excessively and flatten it.
5. Carefully consider the size and orientation of the specimen.
6. Transfer the tissue sample (maximal height 5 mm) onto the stage (Note 8).
7. Immediately apply 2 drops of preparation media onto the sample (sample must not fall dry).
8. Start the tissue chopper. Stabilize the filter paper with forceps during the cutting process, but keep a distance to the moving blade.
9. Transfer the filter paper with the tissue into a new petri dish filled with preparation media.
10. Turn the filter paper upside down just above the medium- the tissue should drop into the medium. If necessary, help with a scalpel by carefully removing the tissue from the paper.

11. If necessary, work with a Stereomicroscope and separate the slices from each other. First use the blunt pipette and pipet up and down to separate the slices. Sometimes a scalpel needs to be used to separate the edges.
12. Transfer each slice to another petri dish for slice collection or use separate petri dishes to control for neighbouring slices, label the lid.
13. Control each slice for the thickness under the Stereomicroscope.
14. A 2 mm or 3 mm biopsy punch is necessary to normalize the tissue size and to get rid of tissue that might have been pinched or quenched while handling the tissue. (Note 9)
15. Group at least 3 slices in one petri dish, forming one condition. (Each condition should be comparable to another; tissue areas can significantly alter from one slice to the other)
16. Transfer 1 ml cultivation medium per well into a labelled 6 well plate.
17. Place each normalized slice with the blunt Pasteur pipette onto a membrane insert, slices are not allowed to be in contact with each other or the edges of the insert.
18. Remove all medium on top of the insert with the tip of a Pasteur pipette or micropipette.
19. Place the insert with the tissue slices in the prepared well plates containing cultivation medium. No air bubbles must be beneath the insert. (Note 10)
20. Transfer the well-plates with the tissue cultures in the incubator (36 °C, 5 % CO_2, standard humidity).
21. Continue with the remaining tissue pieces, if spheroid cultures are needed. (Note 11)
22. Replace medium after 2–4 h to remove excessive accumulation of inflammatory molecules due to tissue damage during cutting. (Note 12)
23. Change the cultivation medium after 24 h, then every 48 h.
24. Start the experiment after 24 h or 48 h. Consider that tissue cultures will adapt to the culture condition, **tumor heterogeneity might get lost after a longer cultivation period of 14 days.** (Note 13)
25. Proceed or store tissue punches according to the endpoint measurements.
    - Soluble secreted markers
      → Store supernatants at −80 °C.
    - Gene expression
      → Store at least 3–4 tissue (3 mm diameter) slices of one condition in 150–200 µl RNA later for 24–72 h at 4 °C prior further processing. (Note 14)

- Histology
→ Rinse tissue punches twice with DPBS.
→ Fix tissue punches at 4 °C fixation buffer (4 % PFA in PBS) for 16–24 h directly on the insert.
→ Store in DPBS at 4 °C.
→ Detach from insert and stain tissue samples or embed in paraffine within 7days.

3.4 Tissue treatment/radiation

- Distribute inserts into different 6-well plates (radiation and sham/controls).
- Sham controls need to be equally transported to avoid misinterpretations of transport artefacts (Fig. 1). This is crucial for OHSC but also important for PDTC. (Note 15)
- Depending on the radiation device, vertical or horizontal adjusting of the tissue is required.

o Horizontal radiation:

Place the 6-well plate into the area of radiation. No further adjustments are necessary. Keep the time out of the incubator as short as possible. Tumor tissue can be handled approx. 2–4 h outside the incubator. However, the medium should be pH stable and the plates should be sealed within a tight clean box or parafilm.

o Vertical radiation:

For vertical beam delivery (required for certain facilities), specialized sample positioning is essential to prevent medium leakage and maintain tissue viability:

1. Design and fabricate 3D-printed brackets with 10° angled intermediate steps to tilt culture plates relative to the vertical beam axis.
2. Position 8 layers of sterile gauze compresses (diameter adjusted to well-size; Hartmann, Heidenheim, Germany; autoclaved before use) beneath each culture insert immediately before irradiation.
3. Fixate the each insert with a lid of a small petri dish to secure positioning within the closed 6-well plate. This configuration prevents membrane drying and medium loss during vertical beam exposure.
4. Seal culture plates with sterile, radiation-transparent film.
5. Position dosimeter alongside samples for verification.
6. Irradiate according to calculated exposure times.
7. To extract the culture medium centrifuge each gauze stack of one condition individually in a falcon tube at 13.000 rpm for 2 min
8. Return cultures to incubator immediately post-irradiation.

Radiation important considerations: (Note 16).
- Verify beam parameters before each session.
- Calibrate X-ray source using ionization chamber dosimetry.
- Monitor beam uniformity using real-time dosimetry.
- Maintain consistent source-to-target distance (typically 50 cm).
- Account for beam degradation through culture medium and plastic.
- Monitor temperature during irradiation (maintain $<25\,°C$).
- Document beam characteristics for each experiment.
- Include non-irradiated controls in identical conditions.
- Accurate dose delivery to thin tissue samples.
- Accounting for medium attenuation and scatter (only particle irradiation).
- Beam homogeneity across sample area.

X-ray Parameters: (Note 17).
- Energy: 320 kVp, 12.5 mA (or equivalent clinical parameters)
- Dose rate: 1.0 Gy/min $\pm$ 0.1 Gy/min
- Dose range: e.g. 0.0, 0.3, 3 Gy
- Field size: Sufficient to cover all culture wells uniformly

Beam Parameters:
- Ion species: e.g. Iron-56 (^{56}Fe)
- Energy: 1000 MeV/u
- Linear energy transfer (LET): ~170 keV/μm
- Dose rate: 1–2 Gy/min
- Dose range: 0.3, 3 Gy

Tissue culture experiment considerations:
- Use appropriate controls to distinguish specific from non-specific effects.
- Combine morphological and functional readouts.
- Include viability assessments alongside activation markers.
- Consider temporal dynamics rather than single time points, if possible.
- Gradual loss of native characteristics over extended culture periods.
- Potential selection for radiation-resistant cell populations.
- Potential confounding effects of prior treatments (clinical specimen).
- Medium composition influences cellular response.

3.5 Analysis

Live cell imaging

- Apply dye or transfection reagents, adeno-associated viral vectors, etc. into the culture medium. Acquire live fluorescence images with a confocal microscope equipped with a temperature- and CO_2 controlled chamber.

- Add PI (1 µg/ml) or any other live dye (Annexin V) to the culture medium, incubate for at least 30 min before imaging- might vary for other dyes. (Note 18)
- AAV application requires an incubation time of 3–5 days to obtain maximal transduction rate
- IsolectinB4 requires 4 h of incubation time, by then microglia and macrophages have taken up the marker and can be visualized for about 48 h.
- Calcein is metabolized very fast, incubation of 20 min is recommended but may vary greatly between patient specimen.

Staining procedure (Note 19).

- Follow the histology instruction point 25 from tissue processing.
- Block with 5 % normal goat serum + 1 % BSA in PBS/Triton 0.5 h (paraffin sections or 4 h whole mount).
- Primary antibodies (overnight or 24 h on shaker, 4 °C, PBS/Triton): Possible makers: Iba1, CD68, P2RY12.
- Secondary antibodies (2 h, room temperature on shaker, PBS/Triton)
- Nuclear counterstain with DAPI.
- Mount and image within 7days.

Quantitative Morphometry (Note 20).

- For whole mount: acquire z-stacks (1 µm step size, 40 × objective), maximum intensity projection for 2D analysis.
- Selection of relevant ROIs and image export from slide-scanner data sets.
- Contrast limited adaptive histogram equalization (CLAHE) for contrast enhancement of somata and processes.
- Detection of somata and processes.
- Identification of individual microglia along with their associated processes.
- Export of individual microglia as small image tiles.
- Selection of 1000 cell image tiles per class (Ramified: High process complexity, small cell body; Activated: Retracted processes, enlarged cell body; Amoeboid: Minimal processes, large rounded cell body) and training of a neural network (CNN).
- Classification of all cell image tiles using the trained network based on microglial morphological phenotypes.
- Morphometric parameter extraction:
  1. Cell body area and perimeter
  2. Convex hull area and perimeter
  3. Cell solidity, convexity and circularity

4. Soma area, perimeter and circularity
5. Process skeleton length, branching points and end points
6. Number of processes, branching index, critical radius, dendritic maximum and Schoenen ramification index (SRI)

4. Notes

1. A blunt Pasteur pipette is ideal to carefully handle the tissue slices. The tip of a Pasteur pipette can be broken off or cut by a diamond engraving pen and use this end for the rubber bulbus.
2. Numerous novel scaffolds are continually being developed, however only a limited number are currently commercially available. Adult tissues from mice and other species can, for example, be cultured on nanotube-based scaffolds (Friebe et al., 2025; Kallendrusch et al., 2017).
3. Other commercial media is also available, e.g. Neurobasal-A medium (Gibco, #10888022) with B-27 Supplement (Gibco, #17504044).
4. The tissue chopper is fast and efficient; however, tissue thickness needs to be carefully controlled. The use of a vibratome is time consuming but provides precision in tissue cutting. Whereas the procedure with the tissue chopper requires only up to one hour until the cultures are prepared for incubation, the procedure with the vibratome lasts often 2–3 h if tissue handling prior cultivation. Whereas the negative examples while handling the tissue with the tissue chopper are given in detail in the text, tissue fixation for vibratome cutting often requires agarose embedding and its success depends on tissue consistency. Tumor tissue often is better chopped as it tends to evade the vibratome blade.
5. Glioblastoma specimen are stable over night at 4 °C, when the transportation tube is only filled with 2–3 ml preparation media. Preferably put it on a roller plate. Do not handle the tissue with force, do not stretch it and do not quench it. Upon arrival, pour the tissue with the media into a petri dish for examination. Carefully inspect the tissue, take care of burns caused by the surgeons, or other handling signs. Remove these parts by cutting the tissue with a scalpel or razor blade. Do not apply force, to stabilize the tissue blunt forceps can be used. Place the cutting razor plate between the forceps and cut the tissue without pressure along the forceps.

6. Cut the tissue into pieces of 5 × 7mm diameter. A long and broad piece will be stable during the cutting process. Often smaller pieces are obtained, then orientation is important to maximize tissue stability during the cutting process.
7. Force and speed of the tissue chopper can be adjusted. At the beginning the speed should be slow to test the tissue, but also to avoid handling troubles. If the speed is too high, the tissue can get smashed. The force should be turned down, if the tissue is very tight.
8. Glioblastoma specimen often adhere very good to the wet filter paper and nothing else should be done. However, very stable (fibrotic) tissues should be glued to the filter paper, also tissues with irregular areas in stiffness. Therefore, put the tissue adhesive onto the filter paper, then dry the tissue on one side and put it on the adhesive. Immediately wet the tissue to prevent the tissue from falling dry. This process has to be done fast.
9. If tissue slices are quenched or pierced or cut during the process, they need to be removed. When all tissue slices are separated, punches are used to normalize the tissue size, start at one edge of the punch and then lower the other side with some force, it is also possible to rotate the punch, push towards the ground to avoid tissue damage. The tissue slices normally fall out easily. When the tissue slice remains in the punch, use an air filled 1 ml syringe and push the tissue slice carefully into the media. Here, we first cut the tissues in slices and then normalize the size of the tissues. This minimizes tissue damage and a narrow control of tissue quality, as each slice is controlled under the microscope
10. Millicell inserts create an air-liquid interface essential for oxygen diffusion and preventing anoxic cell death artifacts. Critical: ensure the membrane surface is completely dry before placing tissue—excess moisture prevents proper membrane contact with medium below, trapping air bubbles. When removing medium from insert top, use blunt micropipettes to avoid direct membrane contact, which causes punctures. Punctured membranes cause medium leakage, destroying the tissue integrity. Check membrane integrity visually under the stereomicroscope after tissue placement.
11. After the tissue cultures are stored in the incubator, the remaining tissue still can be used to obtain tumor spheroid cultures or similar as demonstrated in Fig. 1 described elsewhere (Wen et al., 2023).
12. The first medium replacement 2−4 h post-cutting is important. During the cutting process, enormous quantities of danger-associated molecular

patterns (DAMPs) are released. If this medium remains, it may trigger a cascade of microglial activation and secondary necrotic cell death, independent of any experimental treatment. Subsequent medium changes every 48 h maintain pH and nutrient availability; more frequent changes may stress cells, while less frequent changes allow metabolic acidification.

13. Not all cellular components are yet fully determined. T-cells are present in the tissue and were investigated until day 5 after cultivation. However, each tissue adapts to the culture condition, therefore certain tumor populations may be lost during a prolonged cultivation time.
14. When collecting tissue for downstream molecular analysis of cell death pathways, immediate stabilization with RNAlater is critical. Incubate tissue punches in RNAlater for 24–72 h at 4 °C to allow complete penetration—incomplete penetration creates artifacts in gene expression data. After stabilization, transfer to −80 °C storage. Avoid repeated freeze-thaw cycles, which fragment RNA and compromise quantitative PCR accuracy. Include at least 3–4 independent tissue replicates per condition; biological variability in patient-derived samples often exceeds expected values.
15. Sham controls transported to the irradiation facility but not irradiated represent the critical control for separating actual radiation effects from transport stress-induced activation. Transport controls experience physical movement, temperature fluctuation, and brief oxygen deprivation—all of which activate microglia. Therefore, sham controls need to be identically transported to irradiated samples. Place sham controls in sealed, parafilm-wrapped plates in a temperature-controlled transport box alongside irradiated samples.
16. Energy deposition must be calculated to ensure the Bragg peak occurs exactly within the tissue depth. Request dosimetry verification from your facility's physics team; use ionization chambers positioned at the tissue plane to confirm dose rate and uniformity. Crucially, include immunofluorescence staining for γH2AX (phosphorylated histone H2AX) or 53BP1 as a post-hoc control that tissue received intended radiation dose. DNA damage foci accumulation peaks minutes post-irradiation; image at this timepoint. Absence of expected DNA damage patterns indicates failed irradiation, while excessive off-target damage suggests beam misalignment. This verification step is particularly important for iron-56 beams, where high-LET track structure demands precision (Merz et al., 2013).

17. X-ray irradiation requires ionization chamber dosimetry calibrated to your specific X-ray unit, as dose rate can vary between devices at similar settings. Perform ion chamber measurements prior to your experiments and after any maintenance. Account for dose attenuation through plastic culture plates and medium. Position dosimeters at the tissue plane to measure actual delivered dose, not intended dose. For dose-response curves, use at least 4–5 doses (e.g., 0, 0.3, 1, 3, 5 Gy) to capture the full spectrum of cell death responses.
18. Propidium iodide (PI) enters only cells with compromised membrane integrity (necrotic/late apoptotic cells). Incubate at 5 μg/ml for 2 h; Isolectin B4 (IB4) labels microglia specifically; use over-night incubation as standard.
19. Primary cell death occurs directly in irradiated cells via DNA damage-triggered apoptosis (0.5–6 h). Secondary cell death emerges from microglial phagocytosis and TNF/IL-6-mediated bystander activation (6–48 h). Distinguish these by: (1) co-localizing DNA damage (γH2AX) with apoptotic markers at early timepoints; (2) quantifying cell death distance from irradiated foci— secondary ionization occurs in regions that were not directly hit by the beam trajectory. This distinction is critical for understanding neuroprotection strategies.
20. Machine learning-based microglial morphometry (using convolutional neural networks) provides quantitative phenotyping but demands rigorous training dataset curation. Select at least 1000 cell image tiles per morphological class (ramified, activated, amoeboid), manually validated by at least two independent observers. The number of training tiles per class should be equal. Critical: ensure training datasets include representation from multiple patient specimens and different treatment conditions; a network trained only on control tissue will misclassify activated morphologies in irradiated tissue. After network training, perform post-hoc validation by manually re-examining 100–200 classified cells to assess false positive and false negative rates.

5. Concluding remarks

Understanding radiation-induced cell death is critical for advancing both (1) radiation therapy and (2) the study of the pathophysiological effects of radiation exposure (Nagtegaal et al., 2021). Ionizing radiation induces direct damage to cellular macromolecules and indirect damage through the generation of reactive oxygen species (ROS), ultimately leading to DNA damage (Eriksson & Stigbrand, 2010). This can activate key signalling pathways, including the p53 tumor suppressor pathway and mitochondrial apoptotic mechanisms, which initiate apoptosis, characterized morphologically by chromatin condensation, cell shrinkage, and membrane blebbing (Turnquist et al., 2020). In contrast, necrosis results in uncontrolled cell lysis and triggers a strong inflammatory response, contributing to toxicities in patients (Turnquist et al., 2020). This is observed in conditions such as the Acute Radiation Syndrome (ARS) and various late effects, including tissue fibrosis, vascular injury, and increased risk of secondary malignancies. Cellular senescence and autophagy also play significant roles in these long-term effects (Adjemian et al., 2020; Turnquist et al., 2020). Senescent cells may contribute to chronic inflammation and tissue remodelling, while autophagy can have either protective or cytotoxic outcomes depending on the context. Furthermore, radiation has significant immunomodulatory effects by affecting both innate and adaptive immune cells, which is particularly relevant in the emerging fields of immunogenic cell death (ICD), the abscopal effect, and neuroprotection (Au et al., 2024; Demaria & Formenti, 2020; Rodriguez-Ruiz et al., 2020). The choice of the used experimental platform substantially influences which aspects of radiation-induced cell death can be interrogated. Standard 2D monolayer cultures of isolated microglial or neural cell lines offer excellent accessibility and throughput, enabling rapid screening of candidate neuroprotective agents. However, 2D systems fundamentally lack three-dimensional tissue architecture and native intercellular spatial relationships, limiting investigation of bystander cell death activation and microglial surveillance behaviour in intact networks (Finch et al., 2021; Jo et al., 2024; Warden et al., 2023). Spheroid and organoid systems reconstruct three-dimensional architecture and can maintain cultures for extended periods, but require enzymatic dissociation of the native tissue. In contrast, organotypic slice cultures preserve the native tissue cytoarchitecture, extracellular matrix composition, and functional multicellular interactions within their physiological context. This enables simultaneous investigation of primary cell death (direct radiation-induced apoptosis) and secondary death mechanisms (microglial-mediated bystander activation) without dissociation artifacts. However, this advantage comes with recognized limitations. First, temporal constraints limit investigation to relatively short windows 30–40 days in OHSCs; up to 21–28 days in PDTCs under optimal conditions (Merz et al., 2013; Wen et al., 2023), restricting study of chronic inflammation and late-phase (months post-irradiation) radiation effects. Second, PDTCs exhibit substantially higher biological variability compared to inbred rodent models or standardized cell lines, necessitating larger sample numbers and more complex statistical models to achieve equivalent power. Third, tissue-to-tissue heterogeneity in patient-derived systems, arising from tumor grade, prior treatments, immune infiltration status, and individual genetic background, introduces variables that are scientifically valuable for understanding patient-specific responses but operationally challenging for standardization across multiple institutions (Merz et al., 2010; Miller et al., 2025; Tang et al., 2023). However, recent work demonstrates that astrocytes exhibit substantial divergence between humans and rodents, necessitating caution in interpreting animal data and urges for human model systems to properly understand tissue homeostasis after radiation insults. This pronounced divergence in astrocyte as well as microglial biology between species means that PDTCs and OHSCs are uniquely positioned not for initial drug discovery, but for mechanistic interrogation and validation of preclinical findings, and critically, for understanding tissue homeostasis after radiation insults in a physiologically relevant human context. The ability of PDTCs to recapitulate individual patient tumor biology and treatment responses provides translationally relevant mechanistic insights that cannot be fully captured in more simplistic systems or in vivo rodent models (Gerrits et al., 2020; Li et al., 2021; Pembroke et al., 2021). Although patient-derived tissue cultures face constraints from limited surgical specimen, restricting their use in large-scale pharmacological screening or dose-response curve generation compared to high-throughput platforms, they allow for clinically relevant biomarker studies.

Therefore, developing complex and physiologically relevant model systems is essential for a deeper understanding of radiosensitivity, off-target effects, and the immune consequences of radiation exposure (Finch et al., 2021; Jo et al., 2024; Pangal et al., 2022; Warden et al., 2023). The establishment of patient-derived tissue cultures of surgical resections to select effectiveness and to clarify species differences provides a powerful platform to interrogate microglial biology within its native context. This system allows for the delineation of not only microglial death pathways that reflect and potentially drive tissue damage, offering insight into mechanisms that cannot be fully captured in animal models. Importantly, it enables the identification of optimal therapeutic windows in which microglial activity can be modulated to dampen detrimental inflammation while preserving essential debris clearance. By directly comparing human and murine data, the model provides a critical bridge for validating preclinical findings, thereby advancing the development of effective interventions in neuro-oncology and broader CNS pathology.

Beyond mechanistic studies, these cultures can serve as a clinically relevant testing ground for selected candidate drugs, facilitating biomarker discovery and translational strategies for clinical trial design.

Acknowledgements

The authors thank all clinical staff members for their assistance.

Competing interests

The authors declare no competing interests.

References

Adjemian, S., et al. (2020). Ionizing radiation results in a mixture of cellular outcomes including mitotic catastrophe, senescence, methuosis, and iron-dependent cell death. *Cell Death & Disease, 11*(11), 1003.

Au, N. P. B., et al. (2024). Low-dose ionizing radiation promotes motor recovery and brain rewiring by resolving inflammatory response after brain injury and stroke. *Brain, Behavior and Immunity, 115*, 43–63.

Biswas, K. (2023). Microglia mediated neuroinflammation in neurodegenerative diseases: A review on the cell signaling pathways involved in microglial activation. *Journal of Neuroimmunology, 383*, 578180.

Bourne, J. H., et al. (2025). Microglia are prominent producers of inflammatory cytokines during the hyperacute phase of ischemic stroke. *Communications Biology, 8*(1), 1193.

Carson, M. J., et al. (2006). CNS immune privilege: Hiding in plain sight. *Immunological Reviews, 213*, 48–65.

Constanzo, J., et al. (2021). Radiation-induced immunity and toxicities: The versatility of the cGAS-STING pathway. *Frontiers in Immunology, 12*, 680503.

Daneman, R., & Prat, A. (2015). The blood-brain barrier. *Cold Spring Harbor Perspectives in Biology,* 7(1), a020412.

Demaria, S., & Formenti, S. C. (2020). The abscopal effect 67 years later: From a side story to center stage. *The British Journal of Radiology, 93*(1109), 20200042.

Eriksson, D., & Stigbrand, T. (2010). Radiation-induced cell death mechanisms. *Tumour Biology: the Journal of the International Society for Oncodevelopmental Biology and Medicine, 31*(4), 363–372.

Finch, A., et al. (2021). Advances in research of adult gliomas. *International Journal of Molecular Sciences, 22*(2).

Fogtman, A., et al. (2023). Towards sustainable human space exploration-priorities for radiation research to quantify and mitigate radiation risks. *NPJ Microgravity, 9*(1), 8.

Friebe, S., et al. (2025). Organotypic culture of adult vascularized porcine retina explants in vitro on nanotube scaffolds. *Biological Procedures Online, 27*(1), 35.

Gadani, S. P., et al. (2015). Dealing with danger in the CNS: The response of the immune system to injury. *Neuron, 87*(1), 47–62.

Gerrits, E., et al. (2020). Transcriptional profiling of microglia; current state of the art and future perspectives. *Glia, 68*(4), 740–755.

Gulen, M. F., et al. (2023). cGAS-STING drives ageing-related inflammation and neurodegeneration. *Nature, 620*(7973), 374–380.

Jo, H., et al. (2024). Recapitulating glioma stem cell niches using 3D spheroid models for glioblastoma research. *Biosensors (Basel), 14*(11).

Kallendrusch, S., et al. (2012). Intrinsic up-regulation of 2-AG favors an area specific neuronal survival in different in vitro models of neuronal damage. *PLoS One,* 7(12), e51208.

Kallendrusch, S., et al. (2017). Long-term tissue culture of adult brain and spleen slices on nanostructured scaffolds. *Advanced Healthcare Materials, 6*(9).

Kallendrusch, S., et al. (2013). The G protein-coupled receptor 55 ligand l-alpha-lysophosphatidylinositol exerts microglia-dependent neuroprotection after excitotoxic lesion. *Glia, 61*(11), 1822–1831.

Lawrie, T. A., et al. (2019). Long-term neurocognitive and other side effects of radiotherapy, with or without chemotherapy, for glioma. *Cochrane Database of Systematic Reviews (Online), 8*(8), CD013047.

Leyh, J., et al. (2021). Classification of microglial morphological phenotypes using machine learning. *Frontiers in Cellular Neuroscience, 15*, 701673.

Li, J., et al. (2021). Conservation and divergence of vulnerability and responses to stressors between human and mouse astrocytes. *Nature Communications, 12*(1), 3958.

Merz, F., et al. (2013). Organotypic slice cultures of human glioblastoma reveal different susceptibilities to treatments. *Neuro-Oncology, 15*(6), 670–681.

Merz, F., et al. (2010). Tissue slice cultures from humans or rodents: A new tool to evaluate biological effects of heavy ions. *Radiation and Environmental Biophysics, 49*(3), 457–462.

Miller, T. E., et al. (2025). Programs, origins and immunomodulatory functions of myeloid cells in glioma. *Nature, 640*(8060), 1072–1082.

Mulenge, F., et al. (2024). Transcriptomic analysis unveils bona fide molecular signatures of microglia under conditions of homeostasis and viral encephalitis. *Journal of Neuroinflammation, 21*(1), 203.

Murray, P. J., et al. (2014). Macrophage activation and polarization: Nomenclature and experimental guidelines. *Immunity, 41*(1), 14–20.

Nagtegaal, S. H. J., et al. (2021). Dose-dependent volume loss in subcortical deep grey matter structures after cranial radiotherapy. *Clinical and Translational Radiation Oncology, 26*, 35–41.

Nickl, V., et al. (2022). Glioblastoma-derived three-dimensional ex vivo models to evaluate effects and efficacy of tumor treating fields (TTFields). *Cancers (Basel), 14*(21).

Pangal, D. J., et al. (2022). The abscopal effect: Systematic review in patients with brain and spine metastases. *Neuro-Oncology Advances, 4*(1), vdac132.

Patties, I., et al. (2019). The Chk1 inhibitor SAR-020106 sensitizes human glioblastoma cells to irradiation, to temozolomide, and to decitabine treatment. *Journal of Experimental & Clinical Cancer Research: CR, 38*(1), 420.

Pembroke, W. G., Hartl, C. L., & Geschwind, D. H. (2021). Evolutionary conservation and divergence of the human brain transcriptome. *Genome Biology, 22*(1), 52.

Preissler, J., et al. (2015). Altered microglial phagocytosis in GPR34-deficient mice. *Glia, 63*(2), 206–215.

Prinz, M., et al. (2021). Microglia and central nervous system-associated macrophages-from origin to disease modulation. *Annual Review of Immunology, 39*, 251–277.

Rodriguez-Ruiz, M. E., et al. (2020). Immunological impact of cell death signaling driven by radiation on the tumor microenvironment. *Nature Immunology, 21*(2), 120–134.

Russo, M. V., & McGavern, D. B. (2016). Inflammatory neuroprotection following traumatic brain injury. *Science (New York, N. Y.), 353*(6301), 783–785.

Salem, P. P., et al. (2024). Proton radiation therapy: A systematic review of treatment-related side effects and toxicities. *International Journal of Molecular Sciences, 25*(20).

Schafer, S. T., et al. (2023). An in vivo neuroimmune organoid model to study human microglia phenotypes. *Cell, 186*(10), 2111–2126.e20.

Schwartz, M., & Baruch, K. (2014). The resolution of neuroinflammation in neurodegeneration: leukocyte recruitment via the choroid plexus. *The EMBO Journal, 33*(1), 7–22.
Stoppini, L., Buchs, P. A., & Muller, D. (1991). A simple method for organotypic cultures of nervous tissue. *Journal of Neuroscience Methods, 37*(2), 173–182.
Tang, F., et al. (2023). Tumor-associated macrophage-related strategies for glioma immunotherapy. *NPJ Precision Oncology,* 7(1), 78.
Tinganelli, W., Luoni, F., & Durante, M. (2021). What can space radiation protection learn from radiation oncology? *Life Sciences in Space Research (Amst), 30*, 82–95.
Toprani, S. M., et al. (2024). Cosmic ionizing radiation: A DNA damaging agent that may underly excess cancer in flight crews. *International Journal of Molecular Sciences, 25*(14).
Turnquist, C., Harris, B. T., & Harris, C. C. (2020). Radiation-induced brain injury: Current concepts and therapeutic strategies targeting neuroinflammation. *Neuro-Oncology Advances, 2*(1), vdaa057.
Voshart, D. C., et al. (2024). Radiotherapy induces persistent innate immune reprogramming of microglia into a primed state. *Cell Reports, 43*(2), 113764.
Wang, Y., et al. (2024). Microglia in radiation-induced brain injury: Cellular and molecular mechanisms and therapeutic potential. *CNS Neuroscience & Therapeutics, 30*(6), e14794.
Warden, A. S., et al. (2023). Tools for studying human microglia: In vitro and in vivo strategies. *Brain, Behavior, and Immunity, 107*, 369–382.
Wen, J., et al. (2023). Applications of organoid technology to brain tumors. *CNS Neuroscience & Therapeutics, 29*(10), 2725–2743.
World Medical Association World Medical Association. (2024). Declaration of Helsinki: Ethical principles for medical research involving human participants. *JAMA*. https://doi.org/10.1001/jama.2024.21972.

Printed and bound by CPI Group (UK) Ltd, Croydon, CR0 4YY
20/09/2026
02219559-0001